Medical Device Regulation in the Guangdong-Hong Kong-Macao Greater Bay Area

Medical Device Regulation in the Guangdong-Hong Kong-Macao Greater Bay Area

A Practical Guide for Compliant and Efficient Market Entry

edited by

Jack Wong and
Raymond K. Y. Tong

JENNY STANFORD
PUBLISHING

Published by

Jenny Stanford Publishing Pte. Ltd.
101 Thomson Road
#06-01, United Square
Singapore 307591

Email: editorial@jennystanford.com
Web: www.jennystanford.com

British Library Cataloguing-in-Publication Data
A catalogue record for this book is available from the British Library.

Medical Device Regulation in the Guangdong-Hong Kong-Macao Greater Bay Area: A Practical Guide for Compliant and Efficient Market Entry

ISBN 978-981-5352-38-2 (Hardcover)
ISBN 978-1-042-03128-3 (eBook)

DOI: 10.1201/9781042031283

Contents

Preface

This book supports medical device regulatory initiatives across the Greater Bay Area (GBA): Hong Kong SAR, Macao SAR, and nine mainland cities (Guangzhou, Shenzhen, Zhuhai, Foshan, Huizhou, Dongguan, Zhongshan, Jiangmen, and Zhaoqing). Within a footprint smaller than many single jurisdictions, companies must contend with three distinct legal systems, two special administrative regions, and nine mainland cities—each with its own regulatory logic and market dynamic. The complexity is not merely administrative. It is strategic.

That complexity deserves a practitioner-oriented response. What has been lacking is an integrated, market-entry perspective that treats Hong Kong, Macao, and the GBA mainland cities not as separate problems to be solved in sequence, but as an interconnected system to be navigated as a whole. This book provides it.

The Moment We Are In

Hong Kong stands at a genuine inflection point. The Medical Device Administrative Control System (MDACS) is giving way to a statutory regime under the Chinese Medicine and Medical Products Regulation (CMPR), fundamentally altering compliance obligations for every manufacturer and distributor. The transition will not be uniform. Some product categories will face greater scrutiny; some business models—particularly those relying on informal distributor arrangements—will need to be restructured. Companies that begin preparing now will be in a materially different position from those that wait.

Macao's new Medical Device Law has created a formal regulatory architecture where little existed before. Implementation is ongoing, and practical interpretation continues to be worked out between industry and regulators. This is a period that rewards early positioning and constructive engagement.

The nine mainland GBA cities benefit from targeted policy initiatives, including the Pilot Programme for Hong Kong-Registered

Drugs and Medical Devices in Guangdong—a pathway that did not previously exist. The question is no longer whether these pathways are real, but how to use them effectively within a broader regional strategy that also encompasses the two SARs.

Why This Book, Why Now

When we published our first Asia regulatory handbook in 2013, the pace of change was more manageable. Since then, frameworks across the region have been transformed—new statutory regimes enacted, international standards increasingly adopted, and the GBA emerging as the world's most dynamic bay area economy. The question facing every regulatory affairs professional today is the same: how do we keep pace, and how do we seize the opportunity—including entry into the vast Mainland China market with safe and effective devices that can genuinely benefit its healthcare system?

The editors and contributors share one conviction: regulatory knowledge must be actionable. This book is structured not as a compendium of rules but as a guide to decisions—addressing what each regulation means in practice for submission timelines, distributor relationships, post-market surveillance, and resource allocation.

Part I covers Hong Kong in depth: the MDACS landscape, the CMPR transition, and the steps companies should take now. Particular attention is given to the Licensed Responsible Person (LRP) framework, change management obligations, and designing post-market surveillance systems that are proportionate and audit-ready.

Part II turns to Macao—a market often under-resourced by international players. These chapters address the architecture of the new Medical Device Law and the practical challenges of implementation, drawing on direct experience with the competent authority during the transition period.

Part III examines the GBA as an integrated strategic space. Chapters 7 and 8 argue that Hong Kong and Macao are not merely end-markets but gateway jurisdictions—platforms from which broader mainland GBA access can be structured and de-risked. Chapter 8's comparative analysis is designed to support genuine strategic decisions, not merely catalogue differences.

Part IV closes with a forward-looking chapter on regulatory trends and a 90-day action plan designed to be immediately deployable by regulatory affairs teams regardless of their current state of readiness.

For Whom Is This Book Written

This guide is written for regulatory affairs professionals, market access leaders, and commercial executives responsible for GBA strategy—whether approaching the region for the first time or optimizing an existing presence. It will also benefit legal counsel, distributors, local partners, and academic or policy professionals engaged with GBA medical device regulation.

A Note on Scope and Currency

Regulation in the GBA is not static. We have made every effort to reflect the landscape accurately at the time of writing, distinguishing between provisions in force, those expected but not yet enacted, and areas where direction is clear but destination is not yet fixed. Treat this book as a foundation, not a substitute for current intelligence; references throughout support the ongoing monitoring of developments that will occur after publication.

Acknowledgements

This book is a collective endeavor. The editors are deeply grateful to the chapter authors for sharing hard-won practical knowledge, and to the regulatory professionals, LRPs, distributors, and industry associations across Hong Kong, Macao, and the mainland GBA cities whose candid perspectives have informed every section.

Our thanks also go to the institutions and working groups who patiently explained systems in transition, and to colleagues who offered the critical feedback that only genuine expertise makes possible.

Many thanks to our families (Jack Wong's mother, Shim-Kuen Cheung, wife, Sherry Kwan, and son, Jay Wong; Raymond Tong's parents, Wai-Chuen Tong and Lai-lin Tsui, and children, Lok-Ching, Lok-Tin, and Lok-Ting) for their support, encouragement, and patience. They have been our driving forces.

The GBA is a bold experiment in regulatory convergence and economic integration without precedent. We hope this book serves those navigating it with the rigor, pragmatism, and long-term commitment the opportunity deserves.

Jack Wong
Founder, Asia Regulatory Professionals Association (ARPA)
CEO, RNAscence | CEO, CareerPassion.ai

Raymond Kai yu Tong
Founding Chairman and Professor, Department of Biomedical Engineering, The Chinese University of Hong Kong
Academic Chairman, Hong Kong LRP Panel
Fellow, American Institute for Medical and Biological Engineering

Chapter 1

Support and Participation in the Medical Device Regulatory Initiatives in the Greater Bay Area: Hong Kong SAR, Macao SAR, and Nine Mainland Cities (Guangzhou, Shenzhen, Zhuhai, Foshan, Huizhou, Dongguan, Zhongshan, Jiangmen, and Zhaoqing)

Raymond Kai-yu Tong
[a]*Department of Biomedical Engineering,*
The Chinese University of Hong Kong, Shatin, Hong Kong SAR, China
[b]*Asia Regulatory Professionals Association*
kytong@cuhk.edu.hk

The year 2026 marks a defining moment for Hong Kong and Macao in advancing global medical product regulation. In his opening remarks at the *Medical Products Regulation Symposium* on 3 February 2026, the Director of Health, Dr Ronald Lam, announced the establishment

Medical Device Regulation in the Guangdong-Hong Kong-Macao Greater Bay Area:
A Practical Guide for Compliant and Efficient Market Entry
Edited by Jack Wong and Raymond K. Y. Tong

ISBN 978-981-5352-38-2 (Hardcover), 978-1-042-03128-3 (eBook)
www.jennystanford.com

DOI: 10.1201/9781042031283-1

of the **Hong Kong Centre for Medical Products Regulation (香港藥械監管中心, CMPR)** [1,2]. As outlined in *The Chief Executive's 2025 Policy Address*, the CMPR will implement a phased "primary evaluation" mechanism for new drug registration, empowering Hong Kong to independently assess and approve the safety and efficacy of medical products based on clinical data.

The **primary evaluation mechanism** initially covers extended applications of registered chemical entities—such as new indications, strengths, and dosage forms—laying the foundation for a robust approval system that will later expand to more complex applications. Simultaneously, the government is advancing legislation to establish statutory regulation of medical devices, replacing the voluntary **Medical Device Administrative Control System (醫療器械行政管理制度, MDACS)** [1].

To streamline drug registration, the Department of Health introduced the **"1+" mechanism**, offering one-on-one pre-application consultation meetings [2].

1.1 Macao SAR's 2026 Regulatory Law

In parallel, **Macao SAR** implemented its **Medical Device and Pharmaceutical Products Regulation Law** in 2026, establishing a statutory framework under the **Health Bureau of the Macao SAR (Serviços de Saúde, 澳門衛生局)** [3]. The law harmonises Macao's standards with international norms and introduces mandatory registration for medical devices and pharmaceuticals. This reform strengthens Macao's integration into the Greater Bay Area's regulatory ecosystem, ensuring that products approved in Hong Kong and Macao can be jointly recognised under the **Greater Bay Area (GBA) Connect Scheme (粵港澳大灣區藥械通)** [3,4,5].

1.2 The GBA Connect Scheme

The **GBA Connect Scheme (粵港澳大灣區藥械通)** has become a cornerstone of regional integration and innovation. First introduced in 2021, it allowed five designated hospitals in Guangdong to use Hong Kong- and Macao-approved drugs and devices [4]. By 2023, the second batch expanded coverage to 14 hospitals, and in 2024, the seventh batch brought the total to 45. As of February 2026, **71 hospitals** are designated under the scheme, with **66 drugs and 87 devices** approved for urgent use [5]. These measures directly benefit the **88 million residents** of the GBA, which encompasses Hong Kong SAR, Macao SAR, and nine Guangdong cities—Guangzhou, Shenzhen, Zhuhai, Foshan, Huizhou, Dongguan, Zhongshan, Jiangmen, and Zhaoqing [6].

The scheme's **foundational entry criteria** require that products be approved in Hong Kong or Macao before Mainland China, address unmet medical needs, and undergo review by the **Guangdong Medical Products Administration (**廣東省藥品監督管理局**, GD-MPA)** [5].

1.3 Economic and Research Implications

Beyond healthcare, these developments carry profound economic and research implications. The GBA is emerging as a fertile ground for **BioTech, MedTech, HealthTech, and AgeTech** industries. Leveraging Hong Kong's regulatory expertise and Macao's complementary role, the region fosters cross-border collaboration among academia, industry, and government.

The **map of the GBA**, comprising two SARs and nine Guangdong cities, illustrates the scale of this opportunity. With a population larger than many countries, the region offers unparalleled potential for clinical trials, market expansion, and innovation ecosystems [6].

1.4 Collective Support and Participation

The success of these initiatives depends on active participation from all sectors—government, academia, healthcare institutions, and industry. Guidance documents are regularly updated on official websites, and stakeholders are encouraged to join the **Instrumedica mailing group**, a communication platform managed by the **Department of Health of the Government of the Hong Kong SAR (**香港特別行政區政府衛生署**)** [7].

By engaging through this network and participating in workshops organised by the Department of Health and partner universities, stakeholders can stay informed and contribute to shaping the future of medical device regulation in the Greater Bay Area.

References

1. Department of Health of the Government of the Hong Kong SAR (香港特別行政區政府衛生署). CMPR Preparatory Office. 2026 [cited 2026 May 16]. Available from: www.dh.gov.hk/english/main/main_cmpr/main_cmpr.html
2. Department of Health of the Government of the Hong Kong SAR (香港特別行政區政府衛生署). Medical Products Regulation Symposium. 2026 Feb 3 [cited 2026 May 16]. Available from: www.info.gov.hk/gia/general/202602/03/P2026020300347.htm
3. Health Bureau of the Macao SAR (澳門衛生局). Medical Device Regulation Law. 2026 [cited 2026 May 16]. Available from: www.ssm.gov.mo
4. Drug Office of the Department of Health of the Government of the Hong Kong SAR (香港特別行政區政府衛生署藥物辦公室). Measure of Using HK Registered Drugs/Devices in GBA. 2026 [cited 2026 May 16]. Available from: www.drugoffice.gov.hk/eps/do/en/regulation/gba.html
5. Guangdong Medical Products Administration (廣東省藥品監督管理局, GD-MPA). GBA Connect Scheme Updates (粤港澳大灣區藥械通). 2026 [cited 2026 May 16]. Available from: www.cmde.org.cn
6. Bay Area Development Office of the Government of the Hong Kong SAR (粤港澳大灣區發展辦公室). Medical Services in GBA. 2019 [cited 2026 May 16]. Available from: www.bayarea.gov.hk/en/policy/medical.html
7. Department of Health of the Government of the Hong Kong SAR (香港特別行政區政府衛生署). Instrumedica Mailing Group. 2026 [cited 2026 May 16]. Available from: www.mdd.gov.hk/en/about-us/index.html#subscribe-instrumedica-1

Part I

Hong Kong: Current Practice, Transition, and Preparation

Chapter 2

Current Regulatory Landscape of Medical Devices in Hong Kong

Raymond Kai-yu Tong,[a,b] Tony Yip,[c] Rui Xu,[a] Zhen Wang,[a] Minghao Liu,[a] and Sipeng Zhou[a]

[a]*Department of Biomedical Engineering,*
The Chinese University of Hong Kong, Shatin, Hong Kong SAR, China
[b]*Asia Regulatory Professionals Association*
[c]*Member, Hong Kong Local Responsible Persons Panel*
kytong@cuhk.edu.hk

Hong Kong's regulatory framework for medical devices is undergoing a pivotal transformation as the Government prepares to transition from the long-standing voluntary Medical Device Administrative Control System (MDACS) to a statutory regime [1]. For more than two decades, MDACS has served as the territory's primary mechanism for promoting the effectiveness and safe use of medical devices, providing guidance on classification, listing, and post-market vigilance. Although voluntary, MDACS has shaped industry expectations and established a structured pathway for manufacturers, importers, and distributors seeking to introduce medical technologies into Hong Kong. At the same time, depending

Medical Device Regulation in the Guangdong-Hong Kong-Macao Greater Bay Area:
A Practical Guide for Compliant and Efficient Market Entry
Edited by Jack Wong and Raymond K. Y. Tong

ISBN 978-981-5352-38-2 (Hardcover), 978-1-042-03128-3 (eBook)
www.jennystanford.com

DOI: 10.1201/9781042031283-3

on a device's nature, characteristics, and claims, several existing ordinances may apply—such as the Pharmacy and Poisons Ordinance (Cap. 138), Radiation Ordinance (Cap. 303), Telecommunications Ordinance (Cap. 106), and the Undesirable Medical Advertisements Ordinance (Cap. 231)—creating a multi-layered regulatory environment that newcomers must navigate with care [1].

Recognising the increasing complexity of modern medical technologies and the need for stronger public health protections, the Hong Kong Government has committed to establishing a statutory regulatory framework. Preparatory work is underway, and a bill is expected to be submitted to the Legislative Council in 2026-27, marking a significant shift toward mandatory oversight of device safety, effectiveness and quality management [1]. This transition aligns Hong Kong more closely with international and Mainland China regulatory practices, particularly those of the International Medical Device Regulators Forum (IMDRF), and signals to global manufacturers that compliance expectations will soon become more consistent.

For students, early-career professionals, and companies new to the Hong Kong market, understanding MDACS remains essential—not only because it is the current system, but because it forms the conceptual foundation of the upcoming statutory framework [2]. Listing a medical device in Hong Kong is free of charge so that manufacturers can become familiar with the system and requirement. MDACS introduces the core principles of device classification, conformity assessment, and post-market responsibilities. It also provides an online Medical Device Classification System, a user-friendly digital tool that guides applicants through determining whether their product qualifies as a medical device and, if so, which risk class (Class I–IV) it falls under [2]. This system is particularly valuable for newcomers, as accurate classification is the first and most critical step in preparing technical documentation, identifying applicable standards, and selecting the appropriate conformity assessment route.

Entering the Hong Kong market requires more than regulatory knowledge—it demands strategic preparation. Manufacturers must ensure that their product claims are scientifically substantiated, that their quality management systems align with international standards (such as ISO 13485), and that their documentation is complete and consistent with MDACS requirements [2]. Local Responsible Persons (LRPs) play a central role in device listing, vigilance reporting, and communication with the Medical Device Division (MDD), making early engagement with a qualified LRP essential for smooth market entry [2]. As Hong Kong moves toward statutory regulation, companies that proactively align with MDACS today will be better positioned to meet tomorrow's legal obligations with minimal disruption.

All information in this chapter is provided for reference and experience sharing only. All content should be referred back to the Medical Device Division and the Hong Kong SAR Government for the purpose of medical device control. When submitting the materials must be aligned with the latest control standards in Hong Kong.

2.1 Step-by-Step Guide for Newcomers Entering the Hong Kong Market

To help students, early-career professionals, and companies navigate Hong Kong's regulatory environment, the following structured pathway outlines the essential steps from initial product concept to successful market entry. While the MDACS currently operates as a voluntary scheme, it establishes the core principles that will underpin the mandatory statutory regime anticipated to be introduced in 2026–2027 [2]. Engaging proactively with MDACS positions a newcomer for seamless transition once the legislation takes effect.

First, determine if the product meets the MDACS definition of a medical device, which aligns with the Global Harmonization Task Force (GHTF)/IMDRF definition [2]. The key consideration is whether the product is intended by the manufacturer for a specific medical purpose (e.g., diagnosis, prevention, monitoring, treatment of disease, compensation for injury, investigation/replacement of anatomy, control of conception) and whether it achieves its primary intended action by non-pharmacological, non-immunological, and non-metabolic means. Accessories to an already-defined medical device are also regulated as devices. Products that incorporate human tissues, certain borderline products, or those subject to other legislation (e.g., radiation, pharmacy, or telecommunications ordinances) may fall outside MDACS or require additional approvals. The MDD website provides **guidance document GN-00** (Guidance Notes for Definitions and Abbreviations for Medical Device

Administrative Control System) and the online Medical Device Classification System, an interactive tool that helps newcomers systematically confirm device status [2].

Medical devices are divided into four risk classes for general devices (Class I – lowest risk to Class IV – highest risk) and four classes for in-vitro diagnostic medical devices (IVDMDs: Class A – lowest risk to Class D – highest risk). Classification follows the rules set out in **Technical References TR-003 (GMD) and TR-006 (IVDMD)** [2], which consider factors such as duration of contact, degree of invasiveness, active/non-active nature, and intended purpose. The MDACS online classification tool guides applicants through these hierarchical rules and significantly reduces the chance of misclassification. Correct classification is essential because it dictates the conformity assessment route, the depth of documentation required, and, in the future, the level of regulatory scrutiny. A general rule of thumb: if multiple rules apply, the class reflecting higher risk prevails.

All medical devices placed on the Hong Kong market must be linked to an LRP established in Hong Kong. The LRP acts as the regulatory interface—submitting listing applications, maintaining technical documentation, managing post-market surveillance, reporting adverse events, and communicating with the MDD on behalf of the manufacturer. For overseas manufacturers, the LRP is the authorised representative; a formal designation letter (template provided in Appendix 2 of **GN-01**) should be submitted. The LRP must comply with the **Code of Practice for Local Responsible Persons (COP-01)**, which mandates documented procedures for record-keeping, complaint handling, recalls, and management of adverse events. Selecting a qualified LRP with a solid understanding of ISO 13485 is a strategic early decision.

The application dossier must demonstrate conformity to the **Essential Principles of Safety and Performance of Medical Devices (TR-004)**. The required content scales with device class but typically includes:

- Device description, intended use, and labelling (in both English and Chinese for certain elements)
- Essential Principles checklist and Declaration of Conformity (TR-004)

- Summary Technical Documentation (STED), as described in TR-002
- Risk management report (per ISO 14971)
- Clinical evaluation report, including literature or investigation data
- Evidence of conformity to appropriate standards (e.g., IEC 60601, ISO 10993, ISO 13485)
- Sterilization validation (if applicable)
- Quality management system certificate (e.g., ISO 13485) for the manufacturer
- Labelling and instructions for use in accordance with TR-005

It is strongly recommended to compile documentation following the ASEAN Common Submission Dossier Template (CSDT) structure, as this harmonized format is recognised and facilitates future submissions in the region.

MDACS provides two pathways to demonstrate conformity:

1. **Certification by a Recognised Conformity Assessment Body (CAB):** Currently, BSI, SGS, and TÜV SÜD are recognised under the CAB Recognition Scheme. The CAB conducts an assessment and issues a certificate that attests to the device's compliance with the Essential Principles.
2. **Reliance on marketing approvals from reference jurisdictions:** Valid approvals from specified countries/ regions (e.g., Australia TGA, US FDA, EU notified body, Health Canada, Japan MHLW, Korea MFDS, Singapore HSA) can be used to support the application, supplemented by evidence that the device complies with Hong Kong's specific labelling and safety requirements.

The choice depends on the device's existing approvals and the manufacturer's preferences. For Class I devices, listing is not currently required, but preparing a technical file is still good practice.

Applications for listing Class II–IV general medical devices (using Form MD101) and Class B–D IVDMDs (using Form MD102) are submitted through the online Medical Device Information System (MDIS) portal. The application includes:

- Completed application form with manufacturer, LRP, and device particulars

- Declaration of conformity and essential principles checklist
- Supporting documents (technical file, test reports, clinical evidence)
- Copy of a valid CAB certificate or reference regulatory approval (if applicable)
- LRP documented procedures (e.g., storage temperature control, recall management)

There is currently no charge for listing a medical device. The MDD aims to complete the initial review within about 12 weeks after receiving all necessary information. Once approved, the device receives a Hong Kong Medical Device Listing Number (e.g., HKMD-XXXXX) and is entered into the *List of Medical Devices*.

Once listed, the LRP assumes ongoing responsibilities:

- **Adverse event reporting:** Reportable events (death, serious injury, or public health threat) must be reported to MDD within specified timelines (48 hours for serious public health events, 10 calendar days for death/serious injury, 30 calendar days for other reportable events) using the Adverse Event Report Form.
- **Management of changes:** Any modification to the listed device must be assessed against GN-10 (*Guidance Notes on Changes for Listed Medical Devices*). Depending on the category, the change may require prior approval or simple notification. Unauthorised changes render the listing invalid.
- **Recalls and field safety corrective actions:** LRPs must have documented procedures for product recalls and cooperate fully with the MDD.
- **Listing renewal:** Listings are valid for 5 years. A renewal application must be submitted at least 12 weeks (but not more than one year) before expiry.

The Government's plan to introduce a mandatory registration system in 2026-2027 marks a fundamental shift. The statutory regime will make device registration compulsory, legally enforce LRP obligations, and strengthen post-market controls. Newcomers who voluntarily comply with MDACS today will enjoy a smoother transition: their existing technical documentation, established quality systems, and demonstrated adverse event reporting processes will form a solid foundation for statutory conformity. Early engagement with an LRP, robust classification, and diligent adherence to guidance documents (GN-02, GN-06, TR-003, TR-006) are therefore not only beneficial for immediate market access but also represent a strategic investment in long-term regulatory sustainability.

By following this step-by-step pathway, students, early-career professionals, and companies can systematically navigate Hong Kong's regulatory environment, reduce time-to-market, and build the compliance competence that will be expected under the forthcoming legislation.

2.2 Detailed Explanation of the MDACS Online Classification System

The MDACS Online Medical Device Classification System is a cornerstone tool for any manufacturer, importer, or regulatory professional seeking to introduce a medical device into the Hong Kong market. Its primary function is to provide a structured, transparent, and consistent methodology for determining whether a product falls under the definition of a medical device as defined in the Guidance Notes GN-00 (Definitions and Abbreviations for Medical Device Administrative Control System), and if so, to assign it a precise risk class. The system is a risk management tool that fundamentally shapes the entire regulatory journey of a product, from pre-market conformity assessment to post-market surveillance obligations. Critically, the principles embedded within this online tool are highly consistent with international best practices, most notably those advocated by the IMDRF, and are expected to form the conceptual and technical foundation of Hong Kong's forthcoming statutory regulatory framework.

2.2.1 What the System Does

The online classification tool functions as a digital expert system, guiding users through a logical, rule-based decision tree. The process is fundamentally rooted in a holistic, risk-based philosophy: the level of potential hazard a device poses to patients, users, and the general public under its intended conditions of use dictates the level of regulatory scrutiny it will receive. To achieve a nuanced risk assessment, the system does not rely on a single criterion but evaluates the device across multiple critical dimensions simultaneously.

The first and most crucial dimension is the intended purpose of the device, as declared by the manufacturer. This encompasses a wide spectrum of medical functions, including but not limited to diagnosis, prevention, monitoring, treatment, or alleviation of disease; compensation for an injury or handicap; investigation, replacement, or modification of the anatomy or of a physiological process; and supporting or sustaining life. Closely linked to this is the mode of action. This principle draws a fundamental distinction: a medical device achieves its primary intended purpose through physical or mechanical means (including software, physical barriers, or structural support), not by pharmacological, immunological, or metabolic means. This is the key differentiator that separates medical devices from pharmaceutical products. The system is designed to identify products where a drug action is merely ancillary to a main physical mode of action, anchoring the product firmly within the device regulatory sphere.

Next, the system meticulously evaluates the device's degree of invasiveness and interaction with the human body. This is a graduated scale. A device might be classified as a non-invasive device, one which does not penetrate the body surface or a body orifice. Conversely, it could be an invasive device that enters a body orifice, a surgically invasive device that penetrates the body's surface in a surgical setting, or an implantable device intended to remain in the body after a procedure. The depth and nature of this interaction are primary drivers of risk. Furthermore, the duration of use refines this assessment significantly. The system categorises contact as transient (intended for continuous use for less than 60 minutes), short-term (use between 60 minutes and 30 days), or long-term (use exceeding 30 days). A transient, non-invasive device like a stethoscope presents a vastly different and lower risk profile than a long-term, surgically invasive implant like a hip prosthesis. These two factors—invasiveness and duration—are often evaluated in concert. For example, a transient surgically invasive instrument (like a scalpel) is a lower class than a long-term surgically invasive implant.

Another critical set of factors involves the device's operational technology. The system will ask if the device is an active medical device, defined as one whose operation depends on a source of electrical energy or any source of power other than that directly generated by the human body or gravity. Active devices, such as infusion pumps, electrocardiographs, or surgical lasers, introduce risks related to energy delivery, software control, and potential electromagnetic interference, all of which demand heightened regulatory attention. The system further probes whether an active device is a therapeutic device intended to administer or exchange energy, or a diagnostic device intended to provide information for detecting, diagnosing, or monitoring physiological conditions. The implications of failure are starkly different: a malfunction in a therapeutic radiation unit carries a risk of direct, immediate patient harm, placing it in a higher class than many diagnostic imaging devices. As described in the MDACS Introductory Seminar, the classification rules for active devices (Rules 9 through 12) specifically navigate these distinctions. For instance, a therapeutic device that administers energy in a potentially hazardous way may fall under Rule 9(i), while one that exchanges energy is addressed by Rule 9(ii), and a diagnostic device emitting ionizing radiation triggers Rule 11. A device that is active but does not fall into these higher-risk categories might be captured by Rule 10 or Rule 12, such as a simple active device for administering or removing liquids.

Finally, the system considers special characteristics that introduce unique hazards. This includes whether the device incorporates a medicinal product in an ancillary role (e.g., a drug-eluting stent or an antibiotic-coated catheter), or if it is manufactured from or incorporates non-viable animal or human tissues. Such products are automatically assigned a higher preliminary classification (Rule 14 for medicinal substances, Rule 15 for biological materials) due to the added complexities of pharmacological and biological safety. Devices intended to disinfect or sterilize other medical devices are also treated uniquely under Rule 13. Critically, the system also asks if the device is intended for contraception or the prevention of sexually transmitted disease transmission, triggering a specific classification rule (Rule 16).

Through this iterative questioning, the system systematically narrows down the applicable rules. It is a process of elimination that ensures no single hazard is overlooked. The underlying logic is conservative: where a device could be assigned to multiple classes by more than one rule, the most conservative pathway is always taken, and the highest resulting risk class prevails. This ensures a safety-oriented outcome, particularly for novel or complex products. Although dedicated software medical devices, cybersecurity and AI-Medical Devices have dedicated Technical References (TR-007 and TR-008), the basic principles of this classification framework, such as a diagnostic AI's intended purpose, also serve as the starting point for determining their regulatory control.

2.2.2 Risk Classes

Hong Kong adopts a four-tier, risk-based classification system aligned with the principles established by the IMDRF. Under this framework, the level of regulatory control applied to a medical device is proportional to the potential risk that the device may pose to patients, users, or the public. The system is intended to ensure that devices associated with greater clinical risks are subject to more stringent regulatory oversight, while lower-risk products can follow comparatively simplified regulatory pathways.

The classification of a device depends not on a single characteristic, but on a combination of factors related to its intended use, design, and interaction with the human body. These considerations include the duration of contact with the body, the degree of invasiveness, whether the device is implantable, whether it administers energy or medicinal substances, and whether it exerts local or systemic biological effects. In general, the greater the level of invasiveness and physiological interaction, the higher the associated risk classification.

For example, devices intended for long-term implantation are usually classified into higher-risk categories because they remain in prolonged contact with tissues or organs and may lead to complications if failure occurs. Similarly, life-supporting or life-sustaining devices, such as ventilators and pacemakers, require strict regulatory control because malfunction may directly threaten patient safety or survival. Devices that deliver medicinal substances or therapeutic energy also tend to be associated with higher risk due to the possibility of overdose, incorrect delivery, or tissue damage.

The duration of device use is another important consideration. Devices intended for transient use generally present lower risks than devices intended for continuous short-term or long-term use. For instance, a surgical instrument used temporarily during a procedure is typically associated with lower regulatory concern than a permanently implanted orthopaedic device. In addition, materials and biological interactions may influence classification. An absorbable suture, for example, may be classified at a higher level than a conventional non-absorbable suture because of its prolonged degradation process and potential biological effects within the body.

The classification process also considers whether the device is active or non-active. Active devices rely on external energy sources, such as electricity or batteries, and may involve automated functions, software control, or continuous monitoring. Because these devices may present additional risks related to system failure, software malfunction, or power interruption, they are often subject to higher levels of regulatory scrutiny.

The four classes range from lowest to highest risk as shown in Table 2.1.

Table 2.1 Classification of devices

Class	Risk Level	Examples
Class I	Low	Stethoscopes, wheelchairs
Class II	Low-Moderate	Infusion pumps, surgical gloves
Class III	Moderate-High	Ventilators, orthopaedic implants
Class IV	High	Pacemakers, implantable defibrillators

Class I devices generally present minimal potential harm to users and are therefore subject to relatively limited regulatory requirements. In contrast, Class IV devices are associated with the highest level of risk and consequently require extensive evidence demonstrating safety, quality, and clinical performance before market entry. These devices are also typically subject to stricter post-market monitoring and vigilance obligations.

The classification rules are established in the MDD guidance documents and technical references, particularly GN-01 and TR-

003. These rules provide a structured framework for determining device classification based on defined risk criteria. Where more than one rule applies to a device, the rule leading to the highest risk classification prevails. This conservative principle is intended to prioritize patient safety and minimize the likelihood of under-classification.

In general, lower-risk devices are subject to simplified regulatory procedures, while higher-risk devices require more extensive conformity assessment, technical documentation, clinical evidence, and post-market surveillance. Consequently, the classification system serves as the foundation for determining the level of regulatory scrutiny applied throughout the entire lifecycle of a medical device, from product development and market authorisation to post-market monitoring and risk management.

2.2.3 Why Classification Matters

Accurate classification is a critical first step in the regulatory process, as it determines the entire compliance pathway for a medical device. The assigned risk class influences not only the level of regulatory control before market entry, but also the obligations imposed throughout the product lifecycle, including manufacturing, distribution, post-market surveillance, and product modification.

Specifically, classification dictates the conformity assessment route, including whether a device may be self-declared by the manufacturer or requires evaluation by a recognised third-party CAB. Lower-risk devices may undergo relatively simplified assessment procedures, whereas higher-risk devices are typically subject to more comprehensive regulatory scrutiny. This may include a detailed review of technical documentation, quality management systems, clinical evidence, and risk management processes.

Classification also determines the type and extent of documentation required for device listing and market authorisation. Depending on the risk class, manufacturers may be required to provide device descriptions, intended use statements, essential principles checklists, risk analysis reports, biocompatibility data, performance testing results, software validation information, and clinical evaluation reports. Higher-risk devices generally require more extensive evidence to demonstrate safety, quality, and performance.

Furthermore, the assigned classification influences whether third-party certification is necessary. For many moderate- and high-risk devices, independent assessment by recognised conformity assessment bodies is required to verify compliance with applicable regulatory and quality standards, such as ISO 13485. Consequently, classification has direct implications for regulatory costs, review timelines, and market-entry strategies.

Beyond pre-market considerations, classification also affects post-market obligations. Higher-risk devices are usually subject to stricter vigilance requirements, including adverse event reporting, field safety corrective actions, product recalls, and continuous post-market performance monitoring. Manufacturers and LRPs are expected to establish procedures for complaint handling, traceability, and risk management throughout the operational life of the device.

Importantly, with the anticipated implementation of statutory medical device regulation in Hong Kong in 2026, classification will play an even more significant role in defining legal responsibilities, compliance requirements, and enforcement mechanisms. Regulatory authorities are expected to impose mandatory obligations and stronger enforcement measures, particularly for higher-risk medical devices whose failure may have severe consequences for patient safety and public health.

In this context, the classification system is not merely a technical categorisation tool, but a central element of regulatory strategy that shapes the entire lifecycle management of a medical device. Proper classification is therefore essential not only for regulatory compliance, but also for efficient product development, market access planning, and long-term risk management.

2.3 Comparison Between MDACS and the Upcoming Statutory Framework

As Hong Kong transitions from a voluntary administrative system to a comprehensive statutory regime, it is critical for all stakeholders

– manufacturers, importers, LRPs, and users – to understand the fundamental differences between the current MDACS and the impending mandatory regulatory framework. This chapter outlines the key shifts in legal status, scope, responsibilities, and post-market controls, and discusses what these changes mean for those seeking to place medical devices on the Hong Kong market.

2.3.1 Overview of the Transition

Since its launch in 2004, MDACS has provided a structured yet voluntary pathway for controlling Class II, III and IV general medical devices and Class B, C and D *in vitro* diagnostic medical devices (IVDMDs). While the system has successfully introduced industry to good regulatory practices and raised public awareness of device safety, it lacks the legal force to compel compliance. Depending on a product's nature, certain devices may also be subject to other pieces of legislation (e.g. the Pharmacy and Poisons Ordinance, Cap. 138; the Radiation Ordinance, Cap. 303; the Telecommunications Ordinance, Cap. 106; or the Undesirable Medical Advertisements Ordinance, Cap. 231), creating a fragmented environment.

The Hong Kong Government is now advancing preparatory work for a statutory medical device regulation. A bill is anticipated to be submitted to the Legislative Council in 2026–2027. Once enacted, the regulation will introduce mandatory oversight of the safety, quality and performance of medical devices throughout their life cycle. The forthcoming framework builds upon the principles and infrastructure of MDACS – such as the risk-based classification system, the use of recognised CABs, and the role of the LRP – but elevates many voluntary processes to legal obligations.

2.3.2 Comparison of Key Regulatory Features

Table 2.2 provides a high-level comparison of the current MDACS and the expected statutory framework. Several of these features are then discussed in detail in the following sections.

Table 2.2 Summary of key differences between MDACS and the upcoming statutory framework.

Feature	MDACS (Current)	Statutory Framework (2026, Proposed)
Legal Status	Voluntary administrative system.	Mandatory regulatory regime with statutory powers and penalties.
Device Scope	Class II, III, IV general MDs and Class B, C, D IVDMDs (listing optional).	Likely to mandate pre-market approval for the same risk classes.
Pre-market Control	Voluntary listing or CAB certification recognised; application reviewed by MDD.	Compulsory pre-market authorisation via MDD or designated CABs; established application routes.
LRP Role	Required for voluntary listing; responsible for application, vigilance and recalls.	Legally mandated with defined responsibilities; non-compliance may result in fines or licence suspension.
Quality Management System	Compliance with ISO 13485 or equivalent encouraged but not mandatory for listing.	QMS certification (ISO 13485) expected to be a prerequisite for registration (for manufacturers).
Post-market Surveillance	Adverse event reporting, field safety corrective actions and recalls expected from LRPs of listed devices.	Mandatory reporting for all registered devices and their economic operators; timeframes and formats to be legally binding.
Penalties	None.	Introduction of fines, prosecution, and administrative sanctions for non-compliance.

2.3.3 Legal Status and Scope

The most consequential shift is the move from a voluntary administrative system to a mandatory regulatory regime. Under MDACS, a manufacturer may choose not to list a Class II–IV device and still supply it to the Hong Kong market, provided the product does not fall under other legislation. This will no longer be possible after the statutory regulation enters into force. All Class II, III and IV medical devices (and corresponding IVDMD risk classes) will need to obtain a valid pre-market authorisation before being placed on the market. Class I devices and Class A IVDMDs, which currently are not subject to listing, may continue to be exempted from pre-market approval. The exact scope will be defined in the new regulation.

2.3.4 Legal Status and Scope

MDACS currently provides for listing applications to be submitted by the LRP to the MDD, accompanied by documentation demonstrating conformity with the Essential Principles of Safety and Performance of Medical Devices. Alternatively, a CAB recognised by the MDD may issue a conformity assessment certificate. The statutory framework is expected to formalise these pathways and make pre-market authorisation a legal requirement. It is anticipated that:

- **Recognised CABs** will continue to play a role, but their designation and oversight will be more rigorously structured.
- **Summary Technical Documentation (STED)** and evidence of conformity with internationally recognised standards will remain central to the review process.
- The **online MDIS** will be developed further to handle electronic submissions and tracking.

2.3.5 The Local Responsible Person (LRP)

The role of the LRP is foundational in MDACS and will become even more critical under statutory regulation. In the current voluntary system, an LRP is necessary to list a device and is expected to fulfil various obligations, including maintaining communication with the manufacturer, handling recalls, and reporting adverse events. Under the new law, the LRP will be a **legally accountable entity** with mandatory duties. Key changes include:

- **Legal designation**: The LRP will be the formal representative of the foreign manufacturer, akin to the EU's Authorised Representative. The manufacturer must explicitly appoint the LRP, and the appointment will be registered with the MDD.
- **Expanded responsibilities**: In addition to the current tasks (application submission, record keeping, vigilance), the LRP will be legally responsible for ensuring that the device remains compliant throughout its life cycle, for maintaining a local copy of the technical documentation, and for proactively notifying the MDD of any significant changes or safety issues.
- **Liability and sanctions**: Failure to fulfil LRP obligations can result in administrative penalties, fines, or even suspension of the device registration. It will therefore be essential for manufacturers to select a competent, ISO 13485-aware LRP who can manage the entire regulatory interface.

2.3.6 Quality Management System (QMS)

While MDACS encourages manufacturers to hold ISO 13485 certification, it is not an absolute legal prerequisite for listing. The statutory framework, reflecting international best practices, is likely to require QMS certification (ISO 13485 or an equivalent standard) as a condition for device registration. This will particularly affect local manufacturers and, for overseas manufacturers, those who have not yet sought certification through foreign accredited bodies. The importers' and distributors' obligations will also be reinforced, possibly requiring certification to Good Distribution Practice (GDP) standards.

2.3.7 Post-market Surveillance and Vigilance

Post-market obligations under MDACS – adverse event reporting, management of field safety corrective actions, and recall management – currently rely heavily on the LRP's cooperation for listed devices. The new regulation will make these requirements **mandatory** for all registered devices. Specific changes include:

- **Adverse event reporting**: The timeframes already described in MDACS guidance (e.g. 48 hours for public health risks, 10 calendar days for death/serious injury, 30 calendar days for other reportable events) will be codified in law. Failure to meet these deadlines may incur penalties.
- **Economic operators**: Importers, distributors, and even retailers may be required to report certain incidents or complaints directly to the MDD, expanding the vigilance network.
- **Proactive post-market clinical follow-up**: For higher-risk devices, manufacturers may need to conduct post-market clinical follow-up studies and submit periodic safety update reports, similar to requirements in the EU MDR.

2.3.8 Transitional Arrangements

The Government has indicated that a transitional period will be provided once the statutory regulation comes into effect. During this time, devices that have already been listed under MDACS may be able to transition to the new registration system through a simplified process. Nevertheless, manufacturers and LRPs are strongly advised to begin aligning their documentation, quality systems, and vigilance procedures with the expected requirements well before the legal commencement date.

In summary, the shift from MDACS to a statutory framework represents more than a change in legal status; it signals Hong Kong's full integration into the global regulatory community, with patient safety and product quality at the centre. For newcomers, early preparation and a thorough understanding of the enhanced LRP role, mandatory reporting, and QMS obligations will be key to a smooth and successful market entry.

2.4 Understanding the Local Responsible Person

The LRP is central to Hong Kong's regulatory system.

2.4.1 Who Can Be an LRP?

An LRP may be the authorised representative of the medical device manufacturer, the person responsible for placing the device on the Hong Kong market, or the person responsible for making the application for listing medical devices under the MDACS and bearing the associated responsibilities in relation to the listed devices.

To act as an LRP, the entity should comply with the applicable guidance notes and submit the prescribed application form together with supporting documents, including a valid Business Registration Certificate, documented procedures, and where applicable, a designation letter and quality management system documentation. Comparable documentary and procedural requirements also apply to local manufacturers, importers, and distributors, with variations according to their roles.

From a compliance perspective, the LRP is generally expected to maintain transaction records, ensure appropriate handling and distribution controls, manage product alerts, modifications and recalls, handle complaints, track specific medical devices where necessary, arrange maintenance and servicing, and manage reportable adverse events in Hong Kong. Local manufacturers are expected to perform these functions in accordance with ISO 13485 or equivalent quality management requirements, while importers and distributors are required to maintain procedures appropriate to their activities. Importers additionally bear responsibility for ensuring the standard of medical devices imported.

Local manufacturers, importers, and distributors are also required to make relevant records available for inspection, report adverse events in accordance with Guidance Note GN-03, notify relevant changes, and comply with applicable advertising requirements. For renewal of listing or registration, applications should be submitted at least 12 weeks before expiry; for local manufacturers this is strongly recommended, while for importers and distributors this timing is required.

2.4.1.1 Registration/listing documentation by entity type

Table 2.3 summarizes the principal guidance notes, application forms, and supporting documents applicable to different entities.

Table 2.3 Registration/listing documentation requirements

Item	LRP	Local Manufacturer	Importer	Distributor
Relevant Guidance Notes	GN-01, GN-02	GN-08	GN-07	GN-09
Application Form	MD101 / MD102	LM	MD-IP+D	MD-IP+D
Business Registration Certificate	Required	Required	Required	Required
Documented Procedures	Required	Required	Required	Required
Other Information	1. Designation Letter 2. QMS Certificate, if applicable	1. ISO 13485 certificate or equivalent 2. List of medical devices manufactured	1. List of medical devices imported 2. QMS Certificate, if applicable	1. List of medical devices distributed 2. QMS Certificate, if applicable

2.4.1.2 Core operational and regulatory requirements

Entities participating in the supply and control of medical devices in Hong Kong shall establish and maintain systems appropriate to their functions. The applicable requirements are summarized in Table 2.4.

Table 2.4 Operational requirements by entity type

Requirement	LRP	Local Manufacturer	Importer	Distributor
1. Keeping of transaction records	Applicable	In accordance with ISO 13485 or equivalent requirements	Applicable	Applicable
2. Handling, storage and delivery of medical devices	Applicable	In accordance with ISO 13485 or equivalent requirements	Applicable	Applicable
3. Managing product alerts, modifications and recalls	Applicable	Applicable	Applicable	Applicable
4. Managing reportable adverse events in Hong Kong	Applicable	Applicable	Applicable	Applicable
5. Handling complaints	Applicable	Applicable	Applicable	Applicable
6. Tracking of specific medical devices	Applicable	In accordance with ISO 13485 or equivalent requirements	Applicable	Applicable
7. Arranging maintenance and services	Applicable	In accordance with ISO 13485 or equivalent requirements	Applicable	Applicable
8. Ensuring the standard of medical devices imported	Not applicable	Not applicable	Applicable	Not applicable

2.4.1.3 Compliance, inspection, and reporting obligations

In addition to the above operational requirements, local economic operators are expected to meet the obligations shown in Table 2.5.

Table 2.5 Compliance and reporting obligations

Obligation	Local Manufacturer	Importer	Distributor
Making records available for inspection	Required, including records and documents relating to the QMS or products	Required, e.g. transaction records	Required, e.g. transaction records
Reporting adverse events (see Guidance Note GN-03)	Required	Required	Required
Notifying changes	Required, including any major changes relating to the QMS	Required	Required
Conforming to advertising requirements	Required	Required	Required
Renewal-related arrangement	Suggested to submit renewal application at least 12 weeks before expiry of listing	Required to submit renewal application at least 12 weeks before expiry of listing	Required to submit renewal application at least 12 weeks before expiry of listing

For regulatory and operational purposes, the LRP should be understood as the primary local regulatory contact and accountable party for listed medical devices under the MDACS. While manufacturers, importers, and distributors each bear specific obligations within their own operational scope, the LRP has a broader coordinating role in relation to:

- product listing and maintenance of listing status;
- complaint handling and vigilance coordination;

- adverse event reporting in Hong Kong;
- implementation of product alerts, field safety actions, modifications, and recalls;
- document retention and traceability arrangements; and
- communication with the relevant regulatory authority.

Accordingly, organizations intending to act as an LRP should ensure that they possess:

- formal designation by the manufacturer where applicable;
- documented procedures covering all required regulatory processes;
- sufficient quality management support;
- access to device technical and distribution information; and
- the capability to fulfil post-market responsibilities throughout the device lifecycle.

2.4.2 LRP Responsibilities

The LRP serves as the manufacturer's official regulatory representative within Hong Kong under the MDACS. In this capacity, the LRP assumes a central role in ensuring that the device complies with applicable pre-market and post-market regulatory requirements. The LRP is not merely an administrative contact for filing purposes; rather, it functions as the locally accountable entity responsible for maintaining communication with the MDD, coordinating with the manufacturer and supply chain actors, and supporting the safe and efficacious use of medical devices in the Hong Kong market. The principal responsibilities of the LRP may be summarized as follows.

2.4.2.1 Listing and regulatory communication

One of the core responsibilities of the LRP is to act as the sole eligible applicant for the listing of a medical device under MDACS, except in the case of Class I general medical devices and Class A *in vitro* diagnostic medical devices. The LRP must prepare and submit the listing application to the MDD and ensure that all associated materials, including supporting documentation and labelling samples, are complete, accurate, and properly maintained. In addition, the LRP is responsible for all regulatory communications with the MDD in relation to the application, subsequent clarifications, and any matters arising during the validity of the listing.

2.4.2.2 Documentation, record keeping, and traceability

The LRP is required to maintain appropriate technical and supply records in order to demonstrate compliance and support traceability of devices supplied in Hong Kong. This includes keeping an updated list of importers and distributors, as well as supply records containing, where applicable, information such as the make, model, batch number, serial number, and quantity of devices. The purpose of this obligation is to ensure that devices can be traced efficiently when needed, particularly in the context of complaint investigation, adverse event reporting, or recall. The LRP must also ensure that relevant records and documents are available for inspection and must produce originals or certified true copies to the MDD within the prescribed period upon request.

2.4.2.3 Post-market surveillance, vigilance, and tracking

A further major responsibility of the LRP concerns post-market oversight of listed devices. The LRP must establish mechanisms for monitoring the performance and safety of devices placed on the Hong Kong market, including the handling of complaints, the investigation of adverse events, and the implementation of surveillance activities where required. For specified high-risk devices, such as mechanical heart valves, implantable pacemakers, implantable defibrillators, ventricular support systems, and implantable drug infusion systems, the LRP must maintain a tracking system capable of tracing the device to the patient level where possible, or at least to the user-facility level where patient-level tracking cannot be achieved. In addition, the LRP is required to submit surveillance reports to the MDD at least annually for such devices.

2.4.2.4 Adverse event reporting and corrective actions

The LRP bears direct responsibility for the local handling of reportable adverse events occurring in Hong Kong. Where an adverse event is reported or otherwise comes to the LRP's attention, the LRP must investigate the matter, either independently or in conjunction with the manufacturer or other relevant parties, and report it to the MDD within the applicable timeframe. Events involving death, serious injury, or serious public health concern must be reported as soon as possible and no later than 10 calendar days after the LRP becomes aware of the event, whereas other reportable or potentially reportable events must be reported within 30 calendar days. Closely related to this vigilance function is the LRP's duty to coordinate product recalls, field safety corrective actions (FSCAs), and field safety notices. Upon the issuance of such actions by the manufacturer or overseas authorities, the LRP must inform the MDD promptly, and in any event within 10 calendar days, while also following up with progress reports until the matter has been concluded.

2.4.2.5 Communication, complaint handling, and public interface

The LRP must maintain efficient communication channels with the manufacturer, the MDD, importers, distributors, healthcare institutions, users, and the public. This communication function is fundamental to the operation of MDACS, as it enables updated device information, safety information, and regulatory actions to be transmitted effectively, while also allowing market feedback to be collected and escalated where necessary. As part of this obligation, the LRP must establish a documented complaint-handling procedure and make available appropriate contact details, such as a telephone number, fax number, and/or email address, for the receipt of comments and complaints from users and the public.

2.4.2.6 Advertisement compliance and regulatory liability

In addition to its operational and vigilance-related duties, the LRP must observe certain compliance obligations relating to promotion and legal responsibility. The LRP must ensure that advertisements and promotional materials for the device do not contravene the Undesirable Medical Advertisement Ordinance and do not imply unjustified Government endorsement through improper reference to MDACS listing status. Moreover, the LRP is required to sign the relevant declaration indemnifying the Government against losses and claims arising from its acts or defaults, defective device design, defects in the device, or information supplied to the authorities. In practical terms, this highlights the seriousness of the LRP role and the need for the LRP to consider appropriate risk-control measures, including insurance where necessary.

In summary, the responsibilities of the LRP extend across the full regulatory life cycle of the device, from initial listing and regulatory communication to post-market surveillance, vigilance, traceability, and legal compliance. The LRP therefore serves as the principal local mechanism through which the manufacturer's regulatory obligations are implemented in Hong Kong under the MDACS.

2.4.3 Why the LRP Matters

The LRP occupies a foundational position within Hong Kong's MDACS. Its importance lies in the fact that it provides a legally and operationally identifiable entity within Hong Kong through which the manufacturer's regulatory obligations can be implemented in practice. In a market where many medical devices are manufactured outside the jurisdiction and supplied through importers and distributors, the LRP serves as the essential local mechanism for ensuring continuity of regulatory oversight, effective post-market control, and timely communication with the MDD and other stakeholders. The significance of the LRP may be understood in the following principal respects.

2.4.3.1 Local regulatory accountability

The LRP establishes a clear point of regulatory accountability within Hong Kong. In the absence of a local representative, it would be difficult for the authorities, healthcare institutions, and end users to obtain timely support, communicate safety concerns, or implement corrective actions effectively. By serving as the manufacturer's local regulatory interface, the LRP ensures that there is a responsible entity within the jurisdiction that can respond to regulatory enquiries, provide documentation, and coordinate compliance-related actions.

2.4.3.2 Implementation of lifecycle regulatory obligations

The LRP is significant because it localizes and operationalizes the manufacturer's obligations across the full life cycle of the device. Through the LRP, responsibilities relating to device listing, complaint handling, adverse event reporting, product traceability, recall coordination, and communication with the MDD are translated into concrete and enforceable activities within Hong Kong. This function is particularly important in the post-market phase, where effective oversight depends on the ability to monitor device performance, investigate incidents, and take prompt corrective measures when necessary.

2.4.3.3 Strategic importance for market access and compliance

The appointment of an LRP should not be regarded as a mere administrative formality. For manufacturers seeking entry into the Hong Kong market, the choice of LRP is a strategic compliance decision that may significantly affect the success of listing, the efficiency of regulatory communication, and the quality of post-market oversight. A competent LRP can strengthen the manufacturer's local compliance infrastructure, whereas an unsuitable LRP may create avoidable regulatory and operational risks, including delayed submissions, inadequate complaint management, poor traceability, and ineffective field action execution.

2.4.3.4 Importance of competence and quality management capability

The value of the LRP is closely linked to its practical competence. An effective LRP should have not only knowledge of MDACS requirements, but also sufficient procedural discipline and operational capability to implement those requirements in real market conditions. In this regard, familiarity with quality management principles, including those reflected in ISO 13485, is highly beneficial. Although ISO 13485 certification is not expressly required for acting as an LRP, understanding concepts such as document control, complaint handling, change management, corrective action, and record retention materially enhances the LRP's ability to discharge its responsibilities effectively and consistently.

2.4.3.5 Increasing importance under an evolving regulatory framework

The strategic significance of the LRP is further reinforced by the continued development of Hong Kong's medical device regulatory regime. As MDACS was introduced as an administrative system pending the development of a longer-term statutory framework, the LRP is widely understood as the entity through which local regulatory responsibility is, and is likely increasingly to be, exercised. Its function is therefore comparable in practical terms to that of an in-market regulatory representative in other jurisdictions. For this reason, the LRP should be regarded as a critical component of the manufacturer's regulatory, quality, and risk management infrastructure in Hong Kong.

In summary, the LRP matters because it is the principal local mechanism by which regulatory accountability, post-market control, and practical compliance implementation are achieved under MDACS. Its role is essential not only to market entry, but also to the continued safety, traceability, and regulatory oversight of medical devices placed on the Hong Kong market.

2.4.4 Requirements of LRP

An LRP shall be one of the following:

- a legal person incorporated in Hong Kong; or
- a natural person or legal person holding a valid business registration in Hong Kong.

In addition, the LRP shall be:

- the manufacturer of the medical device; or
- a natural or legal person supported by the manufacturer to perform the obligations of an LRP for the device.

The LRP shall submit the application for listing to the Medical Device Division. The application for listing of the LRP is integrated with the application for listing of the medical device and is made using the same application form.

The LRP shall also establish and maintain documented procedures in accordance with the requirements stipulated by the Medical Device Division.

Either a legal person incorporated in Hong Kong, Or A natural or legal person with business registration in Hong Kong	Either the manufacturer of the device or supported by the manufacturer of the device to perform the obligations of an LRP for the device	Submit the listing application to the Medical Device Division **(The application for listing of LRP is integrated with the application for listing of Medical Devices using the same application form)**	Establish documented procedures according to the requirements stipulated by the Medical Device Division

2.4.5 Established Documented Procedure of LRP

The LRP shall establish and maintain documented procedures and shall submit such procedures together with the application form upon first application for listing.

The documented procedures shall include, where applicable, procedures for:

1. keeping transaction records;
2. handling reportable adverse events in Hong Kong;
3. management of product recalls and field safety notices;
4. control of temperature requirements during storage and transportation, or tracking of specific medical devices;
5. complaint handling; and
6. maintenance and service arrangements.

Such procedures shall be adequate to ensure effective post-market control, traceability, complaint management, incident reporting, and implementation of corrective and preventive actions in relation to the listed medical devices.

The LRP shall ensure that all documented procedures are properly implemented, maintained, reviewed, and updated from time to time.

2.4.6 Relationship between the LRP and the Manufacturer

- **Relationship between LRP and Local Manufacturer**

  Where the manufacturer is a local manufacturer, the following arrangements apply:

- If the local manufacturer acts as the LRP itself, no designation letter is required, since the manufacturer and the LRP are the same legal entity.
- If the local manufacturer appoints another person or company in Hong Kong to act as the LRP, a designation letter must be provided to formally demonstrate the appointment and define the regulatory relationship between the manufacturer and the appointed LRP.

This means that, for a local manufacturer, the need for a designation letter is determined by whether the LRP is:

- the manufacturer itself; or
- a separately appointed Hong Kong-based person or company.

■ **Relationship between LRP and Overseas Manufacturer**

Where the manufacturer is an overseas manufacturer, the following arrangements apply:

- If the overseas manufacturer has a local office in Hong Kong, and the name of the Hong Kong office is the same as that of the overseas manufacturer, the Hong Kong office may act as the LRP itself. In such case, no designation letter is required.
- If the overseas manufacturer has a local office in Hong Kong, but the name of the Hong Kong office is different from that of the overseas manufacturer, the Hong Kong office is regarded as a separate entity for documentation purposes. In this case, a designation letter is required if that Hong Kong office is appointed as the LRP.

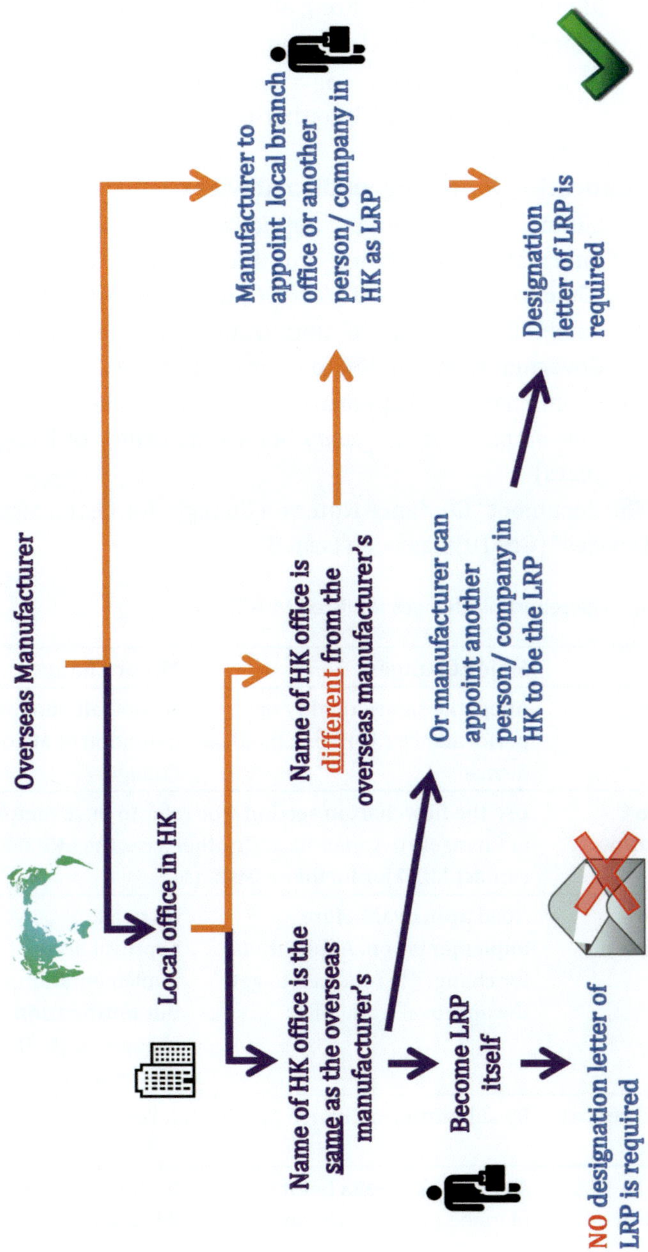
Overseas Manufacturer
Local office in HK
Manufacturer to appoint local branch office or another person/ company in HK as LRP
Name of HK office is the same as the overseas manufacturer's
Name of HK office is different from the overseas manufacturer's
Or manufacturer can appoint another person/ company in HK to be the LRP
Become LRP itself
Designation letter of LRP is required
NO designation letter of LRP is required

- If the overseas manufacturer does not act through an identical-name Hong Kong office, it may appoint:
  - its local office in Hong Kong; or
  - another person or company in Hong Kong to act as the LRP. In either case, a **designation letter** is required.

- **Application for listing medical devices**
  - ☐ Submit the completed application form and required information according to the listing requirements (Guidance Notes GN-02/GN-06) under the MDACS
  - ☐ Establish efficient communication channels with the Government in relation to their application
  - ☐ Submit a renewal application to the MDD at least *12 weeks (but no more than 1 year)* before the expiry of listing (5 years)
- The document "Guidance Notes on Changes for Listed Medical Devices" (GN-10) has been issued.

Table 2.6 LRP reporting changes for listed MDs

	Major Changes	**Minor Changes**
Meaning	Affect the safety, quality or performance (SQP) of a medical device	Do not fall under the definition of Major Change
How to determine	Use the flowchart in section 4 or refer to the Example of Changes in Appendix 1. Or otherwise, the LRP may contact MDD for further assistance.	
How to implement	Need **approval** before implementation. Application for changes is required to get the approval from MDD	No need approval before implementation. But **notification** of changes to MDD is required
How to report or notify	By submitting a Change Application Form	
When to report or notify	**At least 12 weeks** before any planned implementation	Notify MDD **within 24 weeks** from the time the LRP is aware of the change

- GN-10 aims to assist the LRPs in categorising, managing and reporting changes of listed medical devices.
- Starting from 1 January 2024, the LRPs shall comply with the new requirements, and submit the Change Applications with the revised Change Application Form.

2.4.7 Medical Device Adverse Event Reporting

Under the Medical Device Administrative Control System, the Medical Device Adverse Event Reporting System aims to enhance the protection of patients, users, and the public by facilitating the dissemination of safety information that may reduce the likelihood of recurrence of adverse events or mitigate their consequences.

This reporting mechanism is intended for LRPs to report adverse events associated with their listed medical devices where the event has caused, or may cause if repeated, death or serious injury. Submission of an adverse event report should not be interpreted as an admission of liability by the manufacturer, user, or patient, nor does it by itself constitute confirmation that the device failed or caused or contributed to the event.

2.4.7.1 Reportable adverse events

According to the **Guidance Notes for Adverse Event Reporting by Local Responsible Persons (GN-03)**, an adverse event should be reported by the LRP to the MDD when **all** of the following conditions are met:

1. The LRP becomes aware of information regarding an adverse event involving its listed device(s);
2. The LRP's device is associated with the adverse event; and
3. The adverse event has led to one of the following outcomes:
    - death of a patient, user, or other person;
    - serious injury of a patient, user, or other person;
    - no death or serious injury occurred, but the event might lead to death or serious injury if it recurs; or
    - a serious public health concern.

2.4.7.2 Use error

A use error means an act or omission that leads to a result different from that intended by the manufacturer or expected by the operator.

Use errors are reportable in particular where:

- the use error results in death, serious injury, or a serious public health concern;
- the LRP or manufacturer identifies a change in trend or pattern of an issue that may potentially lead to death, serious injury, or a serious public health concern; or
- the LRP or manufacturer initiates corrective action to prevent death, serious injury, or a serious public health concern.

For this purpose, serious public health concern refers to any type of incident that creates an imminent risk of death, serious injury, or serious illness and may require prompt remedial action to prevent a significant risk of substantial harm to the public.

2.4.7.3 Reporting timeframes

The following reporting timelines apply:

- **Within 48 hours**: for an adverse event that has posed, or is likely to pose, a public health risk;
- **Within 10 calendar days**: for adverse events resulting in death or serious injury, counted from the date on which the LRP becomes aware of the incident;
- **Within 30 calendar days**: for all other reportable adverse events, counted from the date on which the LRP becomes aware of the event.

In all cases, reporting should be made **as soon as possible** within the applicable timeframe.

2.4.7.4 Means of reporting

Adverse events may be reported to the MDD by using:

- the **Medical Device Adverse Event Report Form for Local Responsible Persons (Form-Eng AIR-LRP)**; or
- the relevant **electronic reporting form**.

Reports may be submitted through the channels designated by the Medical Device Division, including mail, fax, email, or the web-based reporting platform.

2.5 Classification of General Medical Devices

2.5.1 Introduction to Medical Device Classification

This chapter outlines the principles of medical device classification in accordance with the requirements of the MDACS, adapted from the foundational GHTF guidelines (e.g., GHTF/SG1-N15:2006).

The primary objective of these regulatory controls is to safeguard the health and safety of patients, users, and the general public. By ensuring that manufacturers adhere to specified procedures during the design, manufacture, and marketing phases, regulatory bodies can mitigate potential hazards.

Regulatory controls must be proportional to the level of risk associated with a specific medical device. The stringency of regulatory oversight increases in tandem with the device's degree of risk, always weighing the clinical benefits offered by its use. Crucially, the imposition of these controls is designed to be efficient, avoiding unnecessary administrative burdens on regulators or manufacturers.

2.5.2 General Principles of Risk Assessment

The risk presented by a particular medical device depends substantially on its intended purpose and the effectiveness of the risk management techniques applied during its lifecycle (design, manufacture, and use). Risk is also influenced by the intended user demographic, the device's mode of operation, and the underlying technologies. The classification rules are inherently dynamic, intended to accommodate and regulate emerging medical technologies safely.

2.5.2.1 Factors influencing device classification

A multitude of factors may, either alone or in combination, affect a device's classification. These include:

- **Duration of Contact:** How long the device remains in contact with the human body.
- **Degree of Invasiveness:** Whether the device penetrates the body, and to what extent.

- **Delivery Mechanisms:** Whether the device delivers medicinal products or energy to the patient.
- **Biological Effects:** Whether the device is intended to have a biological effect on the patient, and whether those effects are localized or systemic.

2.5.3 Classification of General Medical Devices

The MDACS classifies general medical devices into four distinct risk classes (Class I, II, III, and IV).

2.5.3.1 Classification rules and guidelines

Manufacturers must document a robust justification for placing their product into a particular risk class. If two or more classification rules apply to a device based on its intended purpose, the device is always allocated the **highest level of classification** indicated.

- **Interdependent Devices:** Where a medical device is intended to be used in conjunction with another medical device (regardless of whether they share a manufacturer), the classification rules apply separately to each device.
- **Assemblages and Kits:** The classification of an assemblage of medical devices that individually comply with all MDACS requirements depends on the manufacturer's overall purpose in packaging and marketing the combination. If one or more devices in the assemblage have not yet met regulatory requirements, the entire combination must be classified as a whole according to its intended use.
- **Accessories:** Accessories intended specifically to enable a "parent" medical device to achieve its intended purpose are subject to the same MDACS requirements as the parent device. However, for classification purposes, an accessory may be classified as a medical device in its own right.
- **Software:** While most software is embedded within a device, standalone software (provided it meets the definition of a medical device) is classified based on its function. If it drives or influences a separate medical device, it is classified according to the intended use of the combination. If independent, it is

classified in its own right. Standalone software is generally deemed to be an *active* device.

2.5.3.2 Risk classes for general medical devices

The four risk classes for general medical devices are illustrated in Table 2.7. Note that these examples are for illustration only; manufacturers must apply the specific classification rules to each device based on its precise claims and intended purpose.

Table 2.7 The four risk classes for general medical devices

Class	Risk Level	Illustrative Device Examples
Class I	Low Risk	Surgical retractors, tongue depressors, simple wound dressings.
Class II	Low-Moderate Risk	Hypodermic needles, suction equipment, non-medicated impregnated gauze.
Class III	Moderate-High Risk	Lung ventilators, bone fixation plates.
Class IV	High Risk	Heart valves, implantable defibrillators.

As the risk class increases, so do the regulatory requirements. Higher-class devices may require extensive technical data submissions, rigorous in-house or independent product testing, frequent independent external audits of the manufacturer's quality system, and independent external reviews of technical data.

2.5.3.3 Example: Non-invasive devices rule (Rule 1)

Classification is highly sensitive to the manufacturer's claims. For example, regarding all non-invasive devices that come into contact with injured skin:

- They are in **Class I** if intended to be used solely as a mechanical barrier, for compression, or for the absorption of exudates (e.g., cotton wool, simple wound dressings healing by primary intent).
- They are in **Class II** if intended to be used principally with wounds that have breached the dermis, including devices intended to manage the microenvironment of a wound (e.g., non-medicated impregnated gauze dressings).

2.5.4 Classification of *In Vitro* Diagnostic Medical Devices

Medical devices used for the *in vitro* examination of specimens derived from the human body are governed by a separate classification matrix. The MDACS categorises IVDMDs into four classes (Class A, B, C, and D) based on both individual risk and public health risk.

2.5.4.1 IVDMD risk matrix

Table 2.8 IVDMD risk matrix

Class	Individual Risk	Public Health Risk
Class A	Low	Low
Class B	Moderate	Low
Class C	High	Moderate
Class D	High	High

2.5.4.2 IVDMD classification rules

- **Rule 1 (Class D - High Public & Individual Risk):** Devices intended to detect the presence of, or exposure to, a transmissible agent in blood, blood components, cells, tissues, or organs to assess suitability for transfusion or transplantation. This also includes devices detecting transmissible agents that cause life-threatening diseases with a high risk of propagation. *(Examples: Tests for HCV, HIV, HTLV).*
- **Rule 2 (Class C, with Class D Exceptions):** Devices used for blood grouping, determining foetomaternal blood group incompatibility, or tissue typing for immunological compatibility.
  - *Class D Exceptions:* Specific critical markers including the ABO system, Rhesus System, Kell System, Kidd System, and Duffy System.
- **Rule 3 (Class C - High Individual Risk):** This broad category includes devices intended for:
  - Detecting sexually transmitted agents (e.g., *Chlamydia trachomatis*).

  - Detecting infectious agents in cerebrospinal fluid or blood with limited propagation risk.
  - Detecting infectious agents where erroneous results could cause death or severe disability (e.g., CMV, MRSA).
  - Prenatal screening for immune status (e.g., Rubella).
  - Human genetic testing (e.g., Cystic Fibrosis).
  - Screening for patient selection as companion diagnostics.
  - Disease staging or monitoring medicines where erroneous results lead to immediate life-threatening situations (e.g., Cardiac markers like Troponin, Prothrombin time).
  - Cancer screening, diagnosis, or staging (e.g., PSA, CEA).
  - Newborn screening for congenital disorders.
- **Rule 4 (Self-Testing / Near-Patient Testing):** Generally classified as **Class C**.
  - *Exceptions:* They fall into Class B (Rule 6) if the result does not determine a critical medical situation (e.g., pregnancy tests, urine test strips). They are elevated to Class D if they fall under Rule 1 or 2 (e.g., rapid self-tests for HIV). A standard Class C example is blood glucose monitoring.
- **Rule 5 (Class A - Low Risk):** General laboratory products or accessories with no critical characteristics, instruments specifically designed for *in vitro* diagnostic procedures, and specimen receptacles. *(Examples: Buffer solutions, histological stains, plain urine cups, ELISA analysers).*
- **Rule 6 (Class B - Moderate Individual Risk):** A catch-all classification for all IVDMDs not covered by Rules 1 through 5. *(Examples: Blood gas analysers, hormone tests, vitamin assays, metabolic markers, HbA1c).*

2.5.5 Conclusion and Compliance

The determination of a device's class is a foundational step in regulatory compliance. Manufacturers must carefully evaluate their product against the definitions of a medical device, document its intended use, and systematically apply the classification rules. Where ambiguity exists, or where a device features characteristics spanning multiple classes, the highest applicable risk class must dictate the conformity assessment pathway.

References

1. Department of Health of the Government of the Hong Kong SAR (香港特別行政區政府衛生署). Medical Device Administrative Control System (MDACS). 2026 [cited 2026 May 16]. Available from: www.mdd.gov.hk/en/mdacs/index.html
2. Department of Health of the Government of the Hong Kong SAR (香港特別行政區政府衛生署). Medical Device Division – Guidance Notes and Technical References. 2026 [cited 2026 May 16]. Available from: www.mdd.gov.hk/en/guidance-notes/index.html

Chapter 3

Preparing for Hong Kong's Statutory Regime and Centre for Medical Products Regulation

Terrenz Leung and Vega Wun
Members, Hong Kong Local Responsible Persons Panel
terrenz.leung@abbott.com

The regulatory information contained in this chapter is intended for market planning and is subject to frequent change. Translations of laws and regulations are unofficial.

3.1 Overview

3.1.1 Centre for Medical Products Regulation (CMPR)

(i) In June 2025, the Department of Health (DH) announced that the CMPR will be established by the end of 2026, and the "primary evaluation" is planned for full implementation by 2030.

Medical Device Regulation in the Guangdong-Hong Kong-Macao Greater Bay Area: A Practical Guide for Compliant and Efficient Market Entry
Edited by Jack Wong and Raymond K. Y. Tong

ISBN 978-981-5352-38-2 (Hardcover), 978-1-042-03128-3 (eBook)
www.jennystanford.com

DOI: 10.1201/9781042031283-4

(ii) The CMPR will unify the regulation of Western medicines, Chinese medicines, and medical devices under the proposed Medical Products Regulation (MPR bill), which is planned to be introduced into the LegCo in 2026.

(iii) The MPR bill will include consequential amendments to the existing Ordinances that regulate pharmaceutical products and Chinese medicine such as the Pharmacy and Poisons Ordinance (Cap. 138).

(iv) The CMPR is aiming to further deepen the collaboration with Mainland national authorities such as the Guangdong Medical Products Administration, GBA Center for Drug Evaluation and Inspection of NMPA, for the enhancement of medical product evaluation in the GBA region.

Apart from the Mainland national authorities, CMPR plans to proactively join the international regulatory organizations. Currently, Hong Kong is actively participating in the events hosted by the World Health Organization (WHO), PIC/S, Forum for the Harmonization of Herbal Medicines, and Global Harmonization Working Party (GHWP). CMPR will strengthen the bonds with these organizations with the goal of becoming an internationally renowned regulatory authority.

3.1.2 Current Medical Device Regulation

Since 2004, the Medical Device Administrative Control System (MDACS) has served as a transitional, voluntary regulatory framework. The Medical Device Division (MDD) under the DH is responsible for the implementation of MDACS.

MDACS adopts risk-based device classification aligned with international IMDRF standards. General Medical Devices (MDs) are classified from Class I to IV, In Vitro Diagnostic Medical Devices (IVDMDs) are classified from Class A to D.

MDACS is regulating medical devices in Hong Kong through pre-market control and post-market surveillance. For the pre-market aspect, the medical device listing covers MDs in Class II or above and IVDMDs in Class B or above. In addition, to encourage the manufacturers and traders to familiarize themselves with the future mandatory requirements, the listing of Local Responsible Person

(LRP), local manufacturer, importer and distributor is also set up as an essential part of the pre-market control.

LRP plays a key role in the medical device listing, especially for effective communication with the manufacturer, importers, users, the public and MDD. LRP is responsible for both pre-market medical device listing and post-market safety alert and adverse event reporting. They are also required to keep the distribution record and ensure the traceability of specific medical devices.

For the post-market aspect, the Medical Device Division proactively collects the safety alerts and recalls from different countries and confirms with LRP the details. Post-marketing data is required to be submitted yearly for high-risk medical devices.

Currently, Hong Kong aligns with global standards and accepts approvals from eight major reference regulators, including the GHTF founding member countries, China NMPA, Singapore HSA, and Korea MFDS. Marketing approvals from at least one of these countries or an MDACS conformity assessment certificate issued by the Conformity Assessment Bodies (CAB) are essential for medical device listing applications.

3.1.3 Moving Toward Mandatory Regulation

Apart from the establishment of the CMPR Preparatory Office and CMPR, MDD is keeping fine-tuning the framework of MDACS for a smoother transition from voluntary to mandatory regulation.

To facilitate the number of voluntary medical device listings under MDACS, DH has implemented new procurement requirements. Starting from 3 March 2026, stage C will be effective, which will mandatorily elevate the bonding between medical device listing and the DH procurement. A new requirement that the MD under purchase must be listed under the MDACS has been added to all quotations and tender invitations issued by DH.

The new procurement strategy issued by DH is also affecting the procurement of Hospital Authorities (HA). Currently, all medical devices applicable for medical device listing that are included in the quotations, and tenders are required to be submitted for listing within 6 months from the effective date of the quotation or tender. Successful tenders/bidders are required to update with HA for the listing record of their MDs under MDACS and/or the status of their submission to DH for listing of their MDs under MDACS from time to time until the MDs are listed.

Aiming to improve the coverage of the MDACS, new technical references (TR) covering new trends of medical device technology, such as SaMD, cybersecurity, AI-MDs, and personalized devices, have been issued by MDD. These technical references provide a clear definition and terminology for the new categories, and detailed guidelines for the technical requirement from MDD.

- TR-007 Software Medical Devices and Cybersecurity
- TR-008 Artificial Intelligence Medical Devices (AI-MD)
- TR-009 Personalized Medical Devices

To boost the efficiency of reviewing medical devices listing applications, the Medical Device Information System (MDIS) launched in 2024 to fully digitize submissions. The MDIS was launched with the new setting so that all enquiries must be answered within 14 calendar days. The review timeline for most of the applications is shortened to within 12 weeks.

3.2 Preparation for the Shifting of Regulatory Environment

3.2.1 Moving Toward Mandatory Regulation

Under the recent transitional environment, it is a must for early preparation for the upcoming legislation process. With the new procurement strategy issued by the DH and HA, products without MDACS listing face commercial exclusion risk. Moreover, Late compliance will become significantly more expensive once statutory enforcement begins. It is important to remember that approval is not the finish line of RA work, but the start of regulatory exposure.

For the companies which rely on unlisted products, a comprehensive submission plan should be drafted as soon as possible to ensure most of the products are listed before the effective date of Medical Product Regulation (MPR). An audit for the portfolio readiness and prioritizing medium- and high-risk devices for full compliance is recommended.

3.2.2 Heightened Requirements for Technical Documentation

As the new technical references are mainly focused on the new digitalized and personalized technology such as SaMD and AI-enabled devices, RA professionals must improve in digital health regulations and AI governance. Technical documentation, especially cybersecurity and AI training/validation data, must be robust. MDD will also focus on the risk management of the software and the AI-related function. Manufacturers must build long-term capabilities in algorithm updates, version control, and ongoing performance monitoring.

To get well prepared, RA professionals must undergo continuous professional development. RA professionals should strengthen their knowledge about software lifecycle management and cybersecurity risk management throughout the whole product lifecycle. Knowledge related to algorithm training, validation, and change control will also help the RA professions to cope with the new trend of medical device technology. RA professions must become technically conversant, not just procedurally competent.

3.2.3 Digitalized Submission Platform Increases Accountability and Transparency

The new MDIS replaces paper-based opacity with a real-time application tracking; the actual status of each submission can be traced on the online platform. Timestamped conversation is replacing messy email communication, increasing the transparency in the well-organized database.

RA workflows must be redesigned for digital execution; the change of recording medium can impact the current procedure. Data protection should be focused on maintaining the quality management system (QMS). Internal alignment between RA, QA, clinical, and engineering is essential to get well prepared for the new style of submission process.

3.2.4 "Stop-Clock" Control of MDIS

MDIS currently does not operate a formal, codified "stop-clock" mechanism like those used by regulators such as FDA, EMA, or Health

Canada. However, a functional stop-clock effect exists in practice whenever the MDD requests additional or clarifying information during the review of an MDACS listing application.

After the launching of MDIS, the time allowed for answering each enquiry is shortened from 3 months to 14 calendar days. The review timeline is effectively paused during this period, which operates as an informal "stop-clock".

To minimize the influence of the new "stop-clock" control, complete, internally validated dossiers shall be prepared for the initial submission. An internal mock deficiency review before MDIS submission is highly recommended. A submission readiness checklist would be helpful to ensure all required documents are ready to be submitted.

In addition, it is useful to gather the questions from the previous submission and prepare the supplementary document before the enquiries are received. For example, the IEC 60601 report is often requested for active medical device, although it is not listed in the guidance notes for submission.

As Hong Kong transitions to a statutory medical device regime under CMPR, the MDIS submission after MPR effect is widely expected to have formalized stop-clock rules with align closer with IMDRF and international review practices. The preparation quality of submission dossier will directly determine time-to-market in Hong Kong.

3.2.5 Regulatory Affairs as a Core Business Function

Once medical devices become regulated, the complexity, visibility, and consequences of regulation have increased significantly. The decision making for the submission strategy will have a revenue impact because of the risk of market access.

RA professions will play a critical role in commercial planning. Broader cross-functional influence would be important to get alignment with the other functions such as marketing, supply chain, and quality assurance.

RA professions would become essential in new product launching, product change management and post-marketing actions. For RA professionals, this transition represents both increased responsibility and elevated strategic importance. Those who adapt early will not only protect compliance but shape competitive advantage.

3.3 Strategic Plan for Regulatory Affairs

3.3.1 Foundation and Gap Closure

To ensure all current products, systems, and processes meet the minimum expectations for a smooth transition into the future statutory regime, it is recommended to undergo a comprehensive review for product portfolio and QMS management.

3.3.2 Portfolio Assessment and Prioritization

To initiate the gap closure, all existing products sold in Hong Kong should be listed. Devices already listed under MDACS should be identified, and it should be ensure that the listing certificates are in good maintenance, such as on-time renewal; all information on the listing certificate are accurate with change management and reporting; and all listing certificates are saved in a secured environment. A complete, validated inventory of all devices supplied in Hong Kong would be important to minimize the gap before the effective implementation of the statutory medical device regime.

It is also important to identify all medium/high-risk devices not yet listed in the business plan to avoid any business impact. If a large number of submissions are required, prioritizing the products critical for revenue or critical for public sector tenders would be helpful.

3.3.3 Strengthen PMS Capabilities

Local Responsible Persons (LRPs) are expected to have an expanded legal obligation under the statutory regime, especially for the post-market surveillance (PMS). To secure a smooth transition, it is critical to review the safety alerts and adverse events reporting procedures and workflows. A more robust complaint-handling system with proactive incident trend analysis will be an effective way to improve and strengthen the PMS capabilities.

3.3.4 Ensure the Digital Submission Readiness

The launch of the MDIS platform is a significant change for the RA environment. To facilitate the change with high compliance to the QMS requirements, it is a must to get alignment for the internal submission formats with the requirement of MDIS. In addition, the importance of cybersecurity awareness and data protection has also significantly increased. A clean digital workflow from dossier preparation to post-marketing surveillance is essential to cope with the new statutory regime.

3.3.5 Talent and Competency Building

To ensure a smooth transition, RA professionals should build up the knowledge of IMDRF principles, AI and software regulations around the world, cybersecurity risk management, and the lifecycle regulation for medical devices. A better understanding of the new trend of medical technology can improve the quality of preparation for the submission dossier.

3.3.6 Regulatory Intelligence and External Alignment

It is foreseeable that CMPR will continuously evolve during and after legislation as benchmarking against FDA, EU MDR, and IMDRF trends. Active monitoring of policy announcements and guidance would be essential to maintain compliance with the regulations. Early scenario planning for the statutory regime by the current guidance notes and the eight reference regulators of listing application is highly recommended.

3.3.7 Regional and Strategic Connections

CMPR positions Hong Kong as a regional regulatory anchor, particularly for the Greater Bay Area through initiatives such as the GBA Measures. The connection between Hong Kong approval, GBA entry approval and the NMPA approval is likely to be deepened. The RA strategy for Hong Kong is foreseeably aligned with China and GBA expansion plans. It would be helpful to understand the recent regulatory environment and strengthen the connections with the related parties such as the distributors.

References

1. https://www.mdd.gov.hk/en/home/index.html
2. https://www.mdd.gov.hk/en/mdacs/index.html#listing-application
3. https://www.dh.gov.hk/english/main/main_pocmpr/main_pocmpr.html
4. https://www.dh.gov.hk/english/main/main_pocmpr/files/cmpr_infobro202506_e_tc.pdf
5. https://www.hkmj.org/abstracts/v31n6/422.htm
6. https://www.mdd.gov.hk/tc/whats-new/measure-of-using-hk-registered-drugs/index.html

Chapter 4

Hong Kong in Practice: Selecting and Managing Local Responsible Persons (LRPs), Compliance Readiness and Experience

Carmen Ling Sephton
Member, Hong Kong Local Responsible Persons Panel
carmen.sephton@gehealthcare.com

4.1 Selecting and Managing Local Responsible Persons (LRPs) Contract Clauses that Support Compliance and Speed

Although Hong Kong's Medical Device Administrative Control System (MDACS) remains a voluntary scheme, listing of medical devices with the Medical Device Division (MDD) of the Department of Health has become a market expectation. Hospitals, clinics, and private healthcare providers increasingly require devices to be listed

Medical Device Regulation in the Guangdong-Hong Kong-Macao Greater Bay Area: A Practical Guide for Compliant and Efficient Market Entry
Edited by Jack Wong and Raymond K. Y. Tong

ISBN 978-981-5352-38-2 (Hardcover), 978-1-042-03128-3 (eBook)
www.jennystanford.com

DOI: 10.1201/9781042031283-5

before procurement, making timely and compliant MDD Listing a commercial necessity.

Furthermore, early listing with the MDD has a whole host of additional advantages. Firstly, the listing comes at no cost during the voluntary phase. For companies with a large portfolio, this will be a major cost saving when the listing transition becomes mandatory. Secondly, the LRP has time to become familiar and keep up to date with the fast-changing guidelines and systems the MDD is utilizing for listing of the medical devices. The MDD currently adopts an online system: the Medical Device Information System (MDIS) for listing submissions. Thirdly, early listing enables a smooth transition to when the regulations become enforced. It is unrealistic to submit the whole product portfolio close to the deadline; to ensure business continuity, it is essential to submit in a phased approach. During this voluntary listing phase, it allows relationship building between both parties. During this early phase of voluntary listing, the MDD often solicits industry or user feedback, and this is an extremely good opportunity to express actual roadblocks or how the industry and the regulators can work hand in hand to safeguard the public use of medical devices.

To be eligible to submit for medical device listing applications to the Hong Kong MDD via the MDACS, the local or overseas manufacturers are required to appoint an LRP. Local manufacturers can act as their own LRP. However, as most medical devices used in Hong Kong are imported from overseas, users or regulators may have difficulty in contacting the manufacturers directly. Thus, the local LRP acts as the bridge of communication between the users, manufacturer, importers, distributors and the HK MDD. The basic criterion for being the LRP is being a legal person incorporated or possessing a business registration in Hong Kong [GN-01]. Serving as the main communication channel, the LRP has also pre- and post-market obligations and regulatory compliance.

4.2 Role of the LRP under the Hong Kong MDACS

Under the MDD Listing of medical devices, the LRP plays a defined and critical role, including:

- Submitting applications for listing and changes to listed devices;
- Acting as the local point of contact with MDD and stakeholders;
- Maintaining records related to listed devices;
- Coordinating complaint handling and adverse event reporting;
- Supporting post-market surveillance activities for listed devices.

While the manufacturer retains ultimate responsibility for device quality, safety and also performance, the LRP's execution directly affects listing timelines and regulatory credibility.

4.2.1 Selecting an LRP: Contract Strategy Starts with Due Diligence

Selecting an LRP for Hong Kong should be treated as a regulatory decision, as the LRP may be the agent that assists the manufacturer transition from the voluntary listing system to when listing migrates to becoming a regulatory obligation.

From a contractual perspective, manufacturers should ensure that the proposed LRP:

- Has prior experience with the MDD Listing process;
- Understands classification rules and technical documentation expectations;
- Can communicate competently with the Department of Health in English and Chinese;
- Has procedures for complaint intake, handling and escalation;
- Is organizationally stable to support continuity;
- ISO13485 or Quality Management System is a plus.

The appointment of the Local Responsible Person should state explicitly that they are appointed for the purpose of MDD Listing and have post listing obligations to ensure expectations of both parties are aligned.

It may be worthwhile to visit the LRP and have a face-to-face meeting prior to the official appointment. This will ensure they have a premises that the MDD recognizes and to ensure they have all the required documentation in place prior to commencement of listing.

4.2.2 Established Quality Management System: ISO 13485 Is a Plus

Possessing certification for ISO 13485 is a plus, but not a mandatory requirement. The ISO 13485 is the international standard that specifies requirements for a Quality Management System (QMS) for organizations involved in the life cycle of medical devices and a regulatory requirement for manufacturers. However, local offices and distributors involved in the distribution and post-market responsibilities can also implement the ISO 13485 QMS. However, medical device expertise as well as personnel resources are required for this to be achieved and may not be feasible for smaller-scale local offices.

If ISO13485 is unavailable, at a minimum, the LRP is expected to maintain the documentation listed in Table 4.1:

Table 4.1

1.	Keeping of supply records	Good record keeping is a must. The supply records are essential to enable tracking of medical device for purposes of product recalls. The supplied batch and lot number, when manufactured and imported, should all be on record. After this, it is also essential to have up-to-date contact information about where specific batches or lots of medical devices have been supplied to. The name, contact phone number, e-mail as well as the address where the product is distributed to should all be on record and easily accessible.
2	Complaints handling procedure	A clear and precise procedure for complaint handling is essential to be able to track and ensure trends do not go amiss. If a particular complaint is seen repeatedly, then this could be perceived as a trend and indicate potential safety issue needs to be investigated and mitigated. This also ensures potential safety issues can be caught at an early stage before actual harm to patients is done.
3	Management of product recalls and field safety notices	When product recalls and field safety notices are issued, the MDD should be alerted for the listed medical device. If the affected products have been distributed in Hong Kong, where they are distributed and the proposed corrections.

4	Handling of reportable adverse events in Hong Kong	Adverse events may be reported by the LRP as well as the user. The corresponding form in the MDIS should be used. The LRP has the responsibility to conduct an in-depth and thorough investigation into the events of the listed medical device and submit a summary report detailing the finding to the MDD.
5	Tracking of specific medical devices	By keeping of supply records, this would also allow the tracking of medical devices. For high-risk medical devices such as implants, it is expected that the tracking is down to the patient level.
6	Maintenance and service arrangements (if applicable)	For some devices with a long product life cycle, there will be an after-sales service for the maintenance and service of the equipment. Typically, there are preventive maintenance checks (PM) and these are scheduled for doing routine checks as the name implies, to ensure the equipment is in good condition and thus preventing any safety issues to arise before they occur. The PM may also include calibration as well as provision of spare parts to when the user requests.

The MDD may at any time request to audit the above records. The responsible LRP should have a systematic record and be able to retrieve the requested information easily.

Perform an annual mock recall audit to ensure the procedures are up to date, adhered to regulatory requirements and reflect actual practices.

4.3 Post-Market Surveillance and Obligations

4.3.1 Product Recall/Safety Notices

The first and foremost responsibility of the LRP is to ensure patients are safeguarded should there be a recall or safety alert issued. The LRP is the communication channel on behalf of the manufacturer, thus responsible for the dissemination of device safety information such as any Field Safety Notice (FSN), both to the customer and the end user plus notification to the MDD. In order to provide the FSN to the end user in a timely manner and subsequently perform the device removal or required corrections of the medical device, the LRP should maintain a list of importers, distributors and supply records of the medical devices. Information such as the model, batch number, serial number and quantity distributed needs to be kept enabling traceability of the medical device. These records are also required to be retained for a period not less than the lifetime of the medical device, or at least as specified by applicable regulatory requirements, whichever is longer.

When product recalls and field safety notices are initiated for listed medical devices, the LRP is expected to notify the MDD with the following information:

- Hong Kong MDD Listing number of the device
- If the affected batch/lot has been distributed in Hong Kong
- Where exactly has the affected batch/lot been distributed and how many of each unit
- The Safey Notice that will be sent to the customer/user
- When the safety notice has been sent
- What corrective actions and preventive actions will be taken for this product recall/safety alert

4.3.2 Adverse Event Reporting

The LRP also needs to ensure they have the resources and ability to comply to the Adverse Event Reporting Guidelines by Local Responsible Persons [Guidance Notes GN-03] as required by the MDACS. It is the LRP's responsibility to report any adverse events occurring in Hong Kong with regard to the listed MD. Timeframes range from 48 hours to 10 or 30 calendar days depending on the category the event falls under. Events that pose a public health risk must be reported within 48 hours, while adverse events that have resulted in a serious injury or death must be reported as soon as possible, but not later than 10 calendar days. For details of definitions of adverse events and previous reporting timelines, please refer to Guidance Notes: GN-03. Upon reporting the adverse event, it is also the responsibility of the LRP to conduct an investigation into the event and report any findings to the MDD as soon as possible. If required, the investigation may be done in conjunction with the manufacturer or other parties.

The LRP should also have a complaint mechanism in place. Complaints may or may not entail safety issues, nonetheless, the LRP should have a telephone number, fax if still used or e-mail address contact provided to the users or to the public should they have comments or complaints.

Tip: The MDD expects the LRP to provide an urgent phone number for 24-hour use.

4.3.3 LRP Qualities

1. Proactive communication and relationship management with MDD
2. Competent Language skills
   Cantonese and English proficient
3. Follow and keep up to date with fast-changing MDACS listing requirements
4. Compliance and ethical conduct
5. Record keeping and inspection readiness
6. Medical device knowledge
7. Possess QMS qualifications
8. Resources

In summary, a good LRP in Hong Kong should be a trusted, knowledgeable while always safeguarding patients' safety first without compromise. A trusted, knowledgeable, and accountable regulatory professional who ensures local compliance, safeguards patient safety, and acts as the key interface between Hong Kong authorities and the manufacturer—while maintaining operational excellence and ethical leadership.

4.4 Change Management Rules (Major/Minor), Timelines, and Planning of Submissions

Medical devices listed under the MDACS in Hong Kong are subject to ongoing obligations to ensure that device information remains accurate, current, and compliant with safety, quality, and performance requirements throughout the product life cycle.

The MDD has published a wide range of Guidance Notes to assist the LRP in listing submissions. The Guidance Notes-10 (GN-10) is for Changes of Listed Medical Devices. Any post-listing modifications should be assessed, classified, and submitted for review in accordance with the (GN-10) [mdd.gov.hk].

Change management under MDACS plays a critical role in safeguarding public health while allowing innovation and continuous improvement of medical devices supplied in the Hong Kong market.

Although MDACS is a voluntary listing system, once a device is listed, in order to maintain listing validity, it is mandatory to be compliant with the MDACS requirements including change notification of the device [mdd.gov.hk].

Guidance Notes for Changes of Listed Medical Devices:

- Principles for managing changes
- Criteria for classifying changes as Major or Minor
- Reporting timelines and approval requirements
- Rules for concurrent supply of original and changed devices [mdd.gov.hk]

Sign up for the newsletter published by the MDD. The Guidance Notes are regularly updated, and it is important to keep abreast of the fast-changing guidelines and requirements.

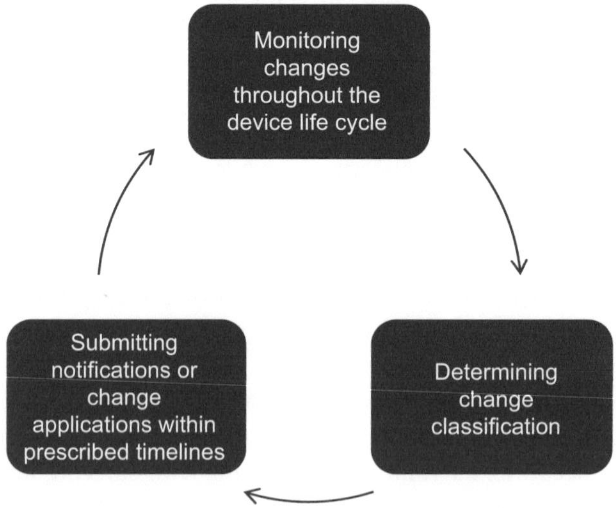

Figure 4.1 Role of the Local Responsible Person (LRP) in maintaining listing validity.

It is important to submit major changes in a timely manner to the MDD for review and approval. Failure by the LRP to comply with GN-10 requirements may result in immediate invalidation of the MDACS listing.

4.4.1 Principles of Change Classification

In line with many established global regulators, the MDD divides changes into two main categories:

- Major changes
- Minor changes

This classification is risk-based, focusing on whether the change could affect the Safety, Quality, or Performance (SQP) of the listed medical device. First and foremost is always the safety of the medical device without compromise. Also, any changes to the QMS would immediately trigger a major change.

Major change can also be coined as 'significant change' or 'substantial change' and is often interchangeable.

4.4.2 Major Changes

A Major Change is defined as any change that could reasonably be expected to affect the SQP of a medical device.

Such changes may:

- Introduce new risks to the patient that are not previously identified
- Increase the probability or severity of existing hazards
- Alter the risk presentation to the user or patient
- Modify intended use or indications

4.4.3 Examples of Major Changes

According to GN-10 and its accompanying flowcharts, examples of major changes include, but are not limited to,

- Design changes affecting form, fit, function, or performance;
- Changes to materials in contact with the patient;
- Changes to sterilization method or sterilization facility;

- Significant software modifications (including cybersecurity or algorithm changes);
- Changes in manufacturing process or critical production environment;
- Labelling changes affecting indications, warnings, or instructions for use;
- Changes to the Quality Management System that may impact product conformity;

Following the GN-10 flowchart allows a step-by-step and easy assessment if the change is in the major or minor category. Once the change category is established, the appropriate supporting documentation should be consolidated and submitted via the MDIS.

For each change category, the MDIS has pre-determined the category. Some categories allow the choice of major or minor change, while some changes as indicated above will automatically be classified as a major change in the form.

4.4.4 Reporting and Approval Requirements

After an initial regulatory impact assessment is made and results in a major change, a list of required change submission documents should be prepared and gathered. Next, a clear timeline should be constructed of when the MDD should be notified. In the instance of a major change, the MDD should be notified as soon as possible and at least 12 weeks before the planned implementation date. Prior approval should be granted by the MDD before implementation.

Supporting documents for major changes:

i. A comparison table of current approved vs proposed changes to show clear differences of before and after
ii. If the changes have been reviewed by the Notified Body, then an approval of the changes would also be required
iii. For labelling changes, a comparison table or a 'red line' version is often requested to show before and after
iv. If the device consists of software, then the software verification and validation report should be submitted
v. Software versioning and software traceability
vi. If the medical device consists of new features reliant on artificial intelligence, then the AI model, data set, performance and evaluation plus deployment should be submitted

4.4.5 Minor Changes

A Minor Change is any change that does not meet the definition of a Major Change, meaning it does not affect the device's safety, quality, or performance.

4.4.6 Examples of Minor Changes

Typical minor changes include:

- Administrative updates (e.g. contact details of manufacturer or LRP)
- Minor editorial labelling corrections with no impact on meaning
- Packaging changes that do not affect sterility or device protection
- Non-critical supplier changes not affecting device conformity

The administrative information of the LRP is pre-determined in the MDIS and thus changes cannot be submitted via the system and should be submitted separately.

4.4.7 Reporting and Approval Requirements

Unlike major changes, minor changes do not require approval from the MDD prior to implementation. However, the LRP is still required to notify the LRP, again via the MDIS Change application within 24 weeks from when the LRP becomes aware of the change.

4.4.8 Determination Process and Flowcharts

GN-10 provides eight structured decision flowcharts to assist LRPs in determining change classification covering the following change principles:

- General changes
- Manufacturing Process and QMS changes
- Design changes
- Sterilization
- Software
- Materials (GMD and IVD)
- Labelling and special listing information

Use of these flowcharts is strongly recommended and serves as a key justification during audits or regulatory scrutiny.

4.4.9 Supply of Changed Medical Devices

Upon approval of the change application, the supplier is allowed to supply both the original and changed versions concurrently. However, clear differentiation between versions must be ensured to avoid confusion among users and healthcare institutions. This allows flexibility and the well planning of stock availability and supply management. A transition plan of timelines should be prepared for the change management. The original devices should only continue to be supplied if they remain compliant with MDACS Essential Principles. The MDD does allow a transition to be completed within 24 weeks.

If this is not achievable and a longer timeline is required, it is pertinent to write and inform the MDD to request a longer transition period and agree upon a new timeline.

4.4.10 Consequences of Non-Compliance

Failure to notify or obtain approval for changes:

- Immediately invalidates the MDACS listing
- Requires cessation of supply
- May negatively impact future renewals or transition to mandatory registration frameworks

4.5 Regulatory Context and Future Outlook

Although MDACS is currently voluntary, the Hong Kong government has clearly signalled a transition toward mandatory medical device registration. Robust change management under GN-10 positions manufacturers and LRPs favourably for future statutory compliance.

The Hong Kong MDACS change management framework reflects international best practices by applying risk-based, pre-emptive regulatory oversight. Proper classification of major and minor changes, timely reporting, and disciplined documentation are essential to maintaining listing status and protecting patient safety. GN-10 should be treated as a core operational regulatory document by any organization supplying listed medical devices in Hong Kong.

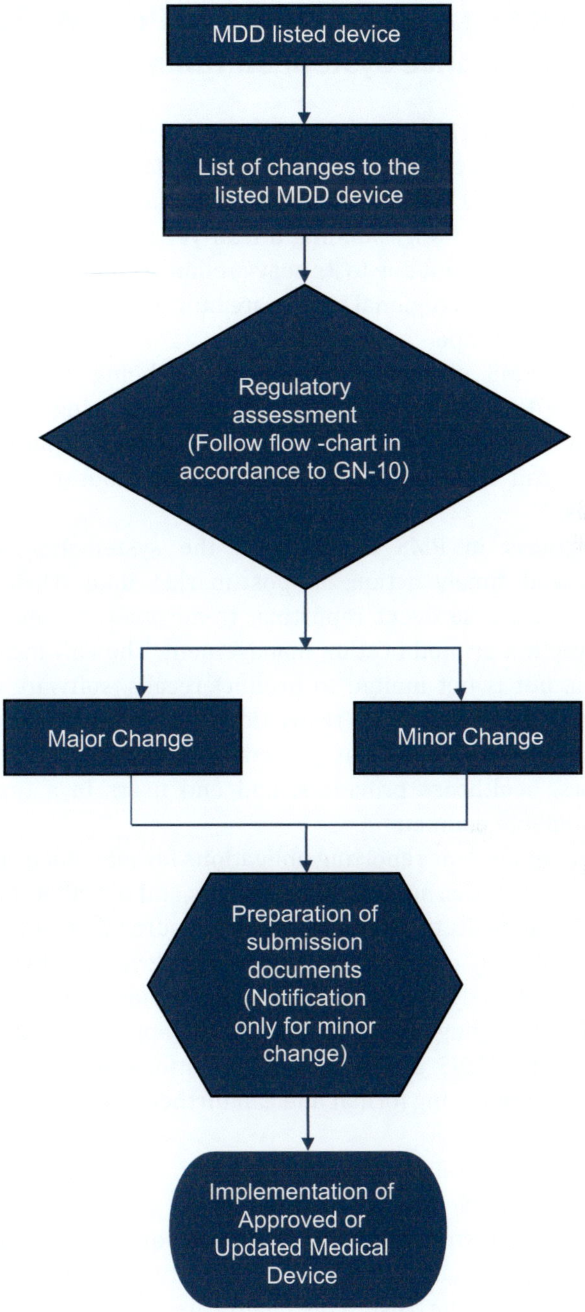

Figure 4.2 Planning of submission of changes of medical devices to the MDD.

4.6 Building a Lean but Robust Post-Market Surveillance System Suitable for Hong Kong

A strong post-market surveillance (PMS) system is essential and a critical pillar in ensuring the continued safety, performance, and quality of medical devices after they are placed and sold on the market. For Hong Kong, building a lean yet robust PMS system is particularly important due to its heavy reliance on imported medical devices, its role as a regional healthcare hub, and its rapidly evolving regulatory landscape.

A lean PMS system should focus on being risk-based. By prioritizing higher-risk medical devices and focusing surveillance resources on products with greater potential impact on patient safety, regulators can achieve effective monitoring without overstretching resources.

Robustness in PMS depends on the systematic collection, analysis, and timely action on post-market data. This includes mandatory adverse event reporting, trend analysis, and effective FSN correction or modification management. The FSN includes the following, but is not limited to product recalls, software updates, labelling changes, and usage restrictions, all of which are critical in ensuring patient safety. Clear procedures must exist for notifying regulators, healthcare providers, and end users in a timely and understandable manner.

Encouraging clear reporting obligations for manufacturers, local responsible persons, authorized importers and distributors, as well as healthcare institutions and end users ensures that safety signals are detected early on and do not go amiss. Beyond individual case reports, trend analysis and periodic safety reviews are essential for identifying emerging risks that may not be apparent from isolated incidents. The MDIS digital reporting platform in place allows a standardized reporting format and can further improve data quality and efficiency.

Stakeholder collaboration is another critical component. A successful PMS system requires close cooperation between regulators, industry, healthcare professionals, and patients. Transparency in regulatory actions and safety communications also helps maintain public trust in the healthcare system.

4.6.1 Preparing for Future Technologies and Evolving Risks

The medical device regulatory landscape is rapidly evolving, particularly with the rise of software as a medical device (SaMD) together with artificial intelligence. These technologies introduce new post-market challenges, including cybersecurity risks, algorithm drift, and continuous learning systems whose performance evolves over time.

A future-ready PMS system must be adaptable, incorporating flexible requirements that allow regulators to respond to novel risks. Continuous monitoring and post-market performance evaluation become increasingly important for these new technologies.

Regular review of PMS policies and performance indicators ensures continuous improvement which is the cornerstone of the quality management system. Feedback loops between post-market findings and pre-market requirements also strengthen the overall regulatory framework.

4.6.2 Conclusion

As medical device technologies evolve, the PMS system must remain adaptable. Regular review of surveillance processes as well as adopting international best practices will ensure that Hong Kong's PMS framework remains both agile and future-ready.

By adopting a risk-based and internationally aligned approach, strengthening vigilance and FSA management, leveraging digital tools, fostering stakeholder collaboration, and preparing for future technologies, Hong Kong can establish a lean and robust post-market surveillance medical device system. This will ensure the benefits of medical devices continue to outweigh their risks throughout their entire lifecycle, safeguarding public health while supporting innovation and regulatory efficiency.

Part II

Macao: New Law and Practical Implementation

Chapter 5

The Medical Device Industry in Macao: Overview of Market Dynamics and Regulatory Frameworks

Hongyu Xian, Junnan Shi, and Carolina Oi Lam Ung
Centre for Pharmaceutical Regulatory Sciences, University of Macau, China
mc56240@connect.um.edu.mo, yc27504@connect.um.edu.mo, carolinaung@um.edu.mo

5.1 An Overview of Macao and the Health Industry

Macao, a special administrative region (SAR) of China, is ranked one of the most densely populated regions in the world, holding a population of 680,000 over an area of 33 km^2. The city also has one of the highest GDP per capita worldwide (surpassing USD 75,500 in 2025). Since its return to China in 1999, Macao has made significant progress in terms of economic and social performance. The GDP increased from USD 6.44 billion in 1999 to USD 51.90 billion in 2025, largely due to the flourishing of the gaming industry.

Medical Device Regulation in the Guangdong-Hong Kong-Macao Greater Bay Area: A Practical Guide for Compliant and Efficient Market Entry
Edited by Jack Wong and Raymond K. Y. Tong

ISBN 978-981-5352-38-2 (Hardcover), 978-1-042-03128-3 (eBook)
www.jennystanford.com

DOI: 10.1201/9781042031283-7

In recent years, to accelerate the diversification of the economy, the Macao SAR government launched the "*Macao Special Administrative Region Plan for Moderate Economic Diversification (2024–2028)*." The initiative outlines a '1+4' strategy, leveraging the strengths of the tourism and leisure industry as a core foundation to drive synergistic growth across four emerging sectors: health, modern financial services, high technology, and the convention, exhibition and commerce industry. As such, the health industry in Macao has now emerged as a priority sector, attracting immense interest and support both locally and beyond.

According to the "*Framework Agreement on Deepening Guangdong-Hong Kong-Macao Cooperation in the Development of the Greater Bay Area*" signed in 2017 by the National Development and Reform Commission and the governments of Guangdong, Hong Kong, and Macao, Macao occupies a key position in the development of the health industry in the Greater Bay Area. Macao's core advantage lies in building an internationally aligned standards system through regulatory innovation in the pharmaceutical and health fields, helping Chinese health products reach the global market.

At the same time, the deep integration of the "*Macao platform + Hengqin space*" provides a strategic hinterland for industrial expansion. As a "super-connector" linking mainland China and Portuguese-speaking countries, Macao fully leverages its role as an economic bridge, effectively promoting regional trade, investment cooperation, and financial services.

5.2 History of Market Architecture and Regulatory Constraints

With technological advancements and social development, medical devices have become a core cornerstone of the modern healthcare system. Their applications are wide-ranging, encompassing various forms from basic consumables and physiological monitoring instruments to high-precision life support systems (such as dialysis machines and ventilators), playing a crucial role in disease prevention, diagnosis, treatment, and rehabilitation. However, Macao has not yet established a dedicated legal framework for medical devices before 2025. Due to a lack of systematic regulatory framework, vague classification standards, and a lack of clinical research and monitoring mechanisms, the research and development, market access, and commercialization of medical devices in Macao have faced significant constraints in the past, which can be exemplified mainly by the following two aspects:

1. Import-Dependent Industry Structure

According to the Statistics and Census Service of Macao, the total import values of medical devices from 2023 to 2025 were $657 million, $835 million, and $679 million MOP respectively (or approximately $81.61 million, $103.73 million, and $84.35 million USD respectively), while the total export values during the same period were only $16.73 million, $27.94 million, and $30.45 million MOP respectively (or approximately $2.08 million, $3.47 million and $3.78 million USD respectively). An import-export scale ratio of 39:1 reflects that Macao's medical device market relies highly on overseas supply with very limited local production capacity.

Moreover, the market supply of medical devices in Macao exhibits a highly concentrated pattern, with the major import sources being the United States, mainland China, and Japan. This singular supply structure signifies a significant path dependence of Macao's medical device supply chain on specific countries or regions, posing considerable risks to the stability of the local market's supply should any supply disruptions occur.

2. Regulatory Gaps

As previously mentioned, before the introduction of the new law, the Macao SAR government did not have relevant specific laws to restrict medical devices. As the medical devices in Macao mainly rely on imports, the government mainly managed medical devices as general import and export goods in accordance with *Law No. 7/2003, "The Foreign Trade Law,"* and classified the manufacturing sites of medical devices as industrial activity sites, subject to the restriction set forth in *Decree-Law No. 11/99/M of March 22nd*. These two laws essentially belong to the categories of trade and industrial management.

In recent years, with the deepening of the medical and health industry and the introduction of high-end medical equipment, the limitations of the traditional, lax regulatory model have become increasingly apparent. This model essentially remains within the scope of "commodity trade regulation," failing to effectively differentiate high-risk medical devices from general commodities, resulting in a lack of differentiated management in safety assessments and quality requirements. Due to the lack of local mandatory registration and approval mechanisms, regulatory authorities struggle to conduct precise pre-market reviews and post-market adverse event monitoring of high-risk products. This regulatory uncertainty has created difficulties for universities, research institutions, and manufacturers in the research and development and compliance processes, becoming a core bottleneck restricting industrial upgrading.

Development Needs

To further implement the Outline Development Plan for the Guangdong-Hong Kong-Macao Greater Bay Area, the National Medical Products Administration and other departments issued the "*Work Plan for Regulatory Innovation and Development of Pharmaceutical and Medical Device in the Guangdong-Hong Kong-Macao Greater Bay Area*" in 2020. The core of this plan is to allow designated healthcare institutions operating in the GBA to use medical devices for urgent clinical purposes used in the public hospitals in Hong Kong and Macao, subject to the approval of Guangdong Province (the Measure). This initiative provides an excellent opportunity for the two regions to build a strategic platform for East-West exchange.

At the same time, the international commercialization of the pharmaceutical industry heavily relies on mutual recognition mechanisms among regulatory agencies. Target exporting countries typically require products to obtain standardized market access permits in their country of origin. Due to the previous lack of a comprehensive legal framework for medical devices in Macao, products developed or distributed in Macao have struggled to obtain internationally credible compliance accreditation. This regulatory vacuum directly weakens the international competitiveness of local products and hinders the compliant path for Macao brands to enter the global market. In light of this, establishing sound institutional regulatory capabilities, particularly in adverse drug reaction monitoring (pharmacovigilance and post-market surveillance), product recalls, and quality traceability throughout the entire distribution process, has become a key to the comprehensive medical device safety management system both regionally and internationally.

5.3 The Establishment of the Macao Pharmaceutical Administration Bureau (ISAF)

Against this backdrop, the Macao Pharmaceutical Administration Bureau (ISAF) was officially established on January 1, 2022, marking the elevation of Macao's pharmaceutical regulatory system from an administrative unit under the Health Bureau to an independent and professional statutory functional agency. The establishment of the ISAF exemplified the endeavor to establish an internationally aligned review system to support the research and development and commercialization of pharmaceutical products. Naturally, the ISAF has become the professional counterpart institution to directly connect with the National Medical Products Administration (NMPA) and international organizations.

The primary task of the ISAF is to fill regulatory gaps, with its legislative focus on the safety, efficacy, and quality assurance of pharmaceutical products. Following the implementation of the *"Law on Pharmaceutical Activities in Traditional Chinese Medicine and Registration of Proprietary Chinese Medicines"* and the optimization of the technical review of new drug registration, the *Medical Device Regulatory System* is scheduled to be formally introduced in 2026 to fill regulatory gaps. For a long time, there has been no specialized legal system in Macao to regulate medical devices and only certain medical device categories were subject to import controls.

To improve the regulatory system, drive industrial innovation, and safeguard public health, Macao is going to enact its first dedicated legislation on medical devices "*Law No. 12/2025—Medical Devices Supervision and Administration System*." This law aims to establish a legal framework covering research and development, production, registration, licensing, market access, and the entire lifecycle of medical devices. Its core lies in establishing a risk-based classification management system, categorizing medical devices into three classes according to their risk level for precise regulation. The new legal framework, while ensuring the quality of local healthcare, lays a solid foundation for Macao to develop into a regulatory innovation for the GBA and even the international pharmaceutical market.

5.4 The Core Framework of the Medical Device Regulatory System

5.4.1 Definition, Mandatory Requirements, and Exemptions

The scope of medical devices is extensive, including a wide range of products used for medical purposes. According to the regulation, a medical device in Macao refers to any instruments, apparatus, appliance, in vitro diagnostic reagents and calibrators, and other similar or related items intended for use directly or indirectly in the human body, including necessary computer software. Their efficacy is primarily achieved through physical means rather than through pharmacological, immunological, or metabolic means, or such means may be involved but only play a supplementary role.

The intended purposes of medical devices include: (1) the diagnosis, prevention, monitoring, treatment, or alleviation of disease; (2) the diagnosis, monitoring, treatment, alleviation, or functional compensation of injury; (3) the examination, replacement, modification, or support of physiological structures or processes; (4) the support or maintenance of life; (5) pregnancy control; and (6) the provision of information for medical or diagnostic purposes through the examination of samples derived from the human body.

The law establishes a regulatory framework centered on two core aspects: "*Medical Device Registration and Filing*" and "*Medical Device Business Activity Management*." The registration or filing of medical devices is a mandatory administrative procedure. Only the medical devices that have been approved for registration or filed may be circulated in the Macao market unless exempted under certain circumstances as listed in Table 5.1. In addition, enterprises engaged in the operation or manufacturing of medical devices must obtain the corresponding licenses in accordance with the law before conducting related business activities in Macao.

Table 5.1 Exemption for the medical device registration and filing

Exempted Cases	Exemption Circumstance
Registration/ Filing Exemption	(1) Medical devices manufactured or imported under the order or approval of the Pharmaceutical Administration Bureau (ISAF) to respond to public health emergencies and device shortages. (2) Medical devices intended exclusively for research and clinical trials. (3) Medical device samples used to form registration dossiers or filing archives. (4) Medical devices exhibited exclusively at academic conferences or exhibition activities. (5) Medical devices imported by public departments and entities to fulfill their duties, approved by the ISAF, to address medical device shortages. (6) Customized medical devices, which are specially designed by practicing physicians according to the patients' medical instructions. (7) Medical devices that have been clinically interpreted by qualified medical practitioners and have been approved by the drug supervision and administration authority and are necessary for the treatment or diagnosis of specific patients' particular conditions.
Operational License Exemption	(1) Entities issued a pharmacy or dispensary license according to Decree-Law No. 58/90/M of September 19 or issued a traditional Chinese medicine pharmacy license according to Law No. 11/2021. (2) Entities issued a license for drug import, export and wholesale operations in accordance with Decree-Law No. 58/90/M of September 19, or issued a license for the import, export and wholesale of traditional Chinese medicine operations in accordance with Law No. 11/2021. (3) The medical device manufacturing plant exports and wholesales the Class IIb and III medical devices manufacture. (4) Public departments and entities importing medical devices to fulfill their duties.

5.4.2 Product Classification, Registration and Filing Procedures, and Requirements

Product classification

The classification rules for medical devices in Macao are very much in alignment with those adopted in Mainland China and the European Union. Medical devices in Macao are classified into Class I, IIa, IIb, and III based on the level of risk (Figure 5.1). The risk level assessment of medical devices includes not only the intended purpose but also structural characteristics and methods of use of the product. Every risk level corresponds to a specific admission pathway. Class I medical devices carry the lowest risk while Class III devices carry the highest.

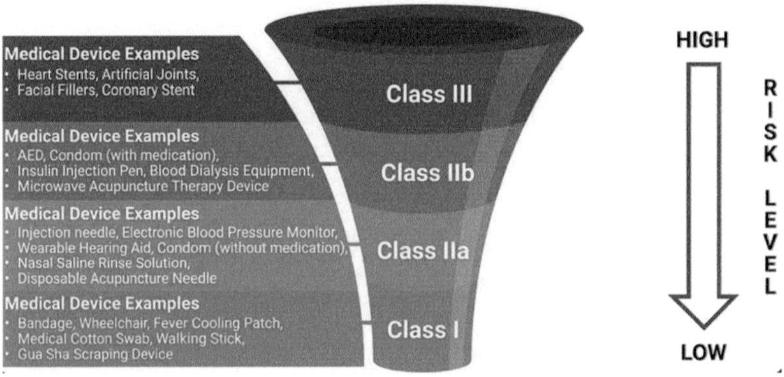

Figure 5.1 Classification of medical devices in Macao.

It is worth noting that if the classification of a medical device is adjusted from Class III to Class II or Class I, or from Class IIb to Class IIa or Class I, the relevant registration certificate for the medical device shall remain valid until its expiration date. If the medical device is adjusted from Class I to Class II or Class III, or from Class IIa to Class IIb, even from Class II to Class III, the registration holder or listing holder must submit either an application for registration change or a registration application or an application for listing change to ISAF within a specified period of time. The above classification rules apply to medical devices used in both traditional Chinese Medicine and Modern Medicine.

Furthermore, medical devices can also be used for non-medical purposes. For example, beauty salons may use certain medical devices for non-medical beauty procedures. To determine whether a product is a medical device, it should be based on whether it conforms to the legal definition of medical devices, rather than whether it is included in the classification directory. That is, the definition takes precedence over the directory. Anything that meets the definition will be subject to regulation, and the drug regulatory authority will also update the medical device classification category in a timely manner.

Registration, filing procedures, and requirements

Law No. 12/2025 distinguishes the registration and filing procedures based on risk levels: Class IIb and Class III are considered high-risk devices and require registration procedures; while Class IIa and Class I have relatively lower risks and can undergo filing procedures only. A comparison of the requirements for medical device registration and filing is illustrated in Table 5.2.

Registration of a medical device refers to the application submitted by the applicant in accordance with the legal procedures and requirements. After the ISAF reviews the safety, effectiveness, and quality controllability of the medical devices, it issues a registration certificate to the applicant. The registration process of medical devices mainly covers three core stages: pre-registration consultation, registration review, and registration inspection:

1. *Pre-registration Consultation*: Applicants may consult ISAF regarding the required documents, declaration materials, standard procedures and fees before formal submission to prepare in advance and improve subsequent review efficiency.
2. *Registration Review*: The registration review is further divided into two levels: the form review and the substantive review, which are carried out sequentially. The ISAF completes the formal review within 20 working days after receiving the application. Once it confirms that the applicant is legitimate and the materials are complete, it initiates the substantive review to verify whether the application materials comply with the relevant regulations and registration requirements of medical devices in Macao. The legal time limit for the

substantive review is calculated from the date when the completed materials are accepted. Due to the different risk levels of the devices, the review timelines vary: 70 working days for Class IIb devices and 100 working days for Class III devices. The longer review period reflects the stricter evaluation standards required for high-risk devices.

Table 5.2 A comparison of the requirements for medical device registration and filing

Feature	Filing	Registration
Applicable Categories	Class I, Class IIa	Class IIb, Class III
Review Mechanism	After submitting the complete file, just notify for confirmation	Requires substantive review and approval by the ISAF and an expert committee
Clinical Evaluation	General exemption, or based on existing literature data	Mandatory clinical evaluation/clinical trials required
Validity Period	Valid long-term	Five years (renewable)
Alteration Rules	Any changes to labels or specifications must be notified in writing five days in advance	Substantive changes must be approved in advance; administrative changes must be notified 20 days in advance
Revocation Risk	Higher—If the review reveals non-compliance or safety issues, the drug regulatory authority may revoke it	High—Registration may be suspended or revoked if there are serious security risks

3. *Inspection*: The ISAF may conduct on-site inspections of the research and manufacturing sites of medical devices to verify the authenticity and consistency of the application materials. The ISAF will notify the applicant at least 10 working days before the inspection date. The applicant may submit an application to adjust the inspection date and provide reasons.

If any deficiencies are found during the inspection, the ISAF may require the applicant to make corrections within the specified period.

Filing of a medical device refers to the act of submitting the relevant information of medical devices to the ISAF in accordance with legal procedures and requirements, so that the ISAF can archive and keep it for future reference. Compared to registration, the filing process is simpler and mainly consists of two stages: pre-procedure consultation and formal review:

1. *Pre-filing Consultation*: The applicant can consult ISAF about the requirements, necessary documents and procedures for the filing before the filing process.
2. *Formal Review*: The ISAF conducts a form review of the submitted materials after the application for registration is submitted, mainly to check whether the applicant is legitimate and whether all relevant documents are complete. Once it is confirmed that the submitted materials for archiving and reference are complete, the ISAF will notify the applicant that the filing process is complete.

It is necessary to ensure that the medical devices meet the corresponding quality standards (including mandatory national standards or other standards recognized by the drug supervision and administration bureau), are effective and safe (not posing any danger to human health under normal use conditions), and have names, labels and instructions that comply with legal and supplementary regulations. If necessary, the ISAF will seek assistance from expert committees or third-party evaluation institutions through consultation, etc., to comprehensively evaluate whether the quality standards of the products are sufficient to ensure that the medical devices meet the requirements of safety, effectiveness and controllable quality.

In addition to the general requirements mentioned above, the law sets specific provisions for medical devices protected by patents. If the applicant is not the owner of the relevant patent, the registration or filing application can only be submitted for 240 days before the patent expires. Even if the application is accepted, the ISAF must wait until the patent officially expires before approving the registration or completing the filing notification. At the procedural level, when applying for medical device registration or filing, the applicant must submit relevant certification documents to the ISAF to explain the patent rights of the product. If the product is not protected by patents, a relevant statement must be submitted so that the ISAF can understand the patent rights status of the product.

5.4.3 Clinical Evaluation (Including Exemptions)

The clinical evaluation of medical devices is a fundamental and crucial step to ensure their safety and effectiveness, and to run through the entire process from before the product's launch to after its release. Essentially, clinical evaluation involves using scientific methods to collect, analyze, and evaluate relevant clinical data, thereby making a comprehensive judgment on the safety, effectiveness, and risk-benefit status of the medical devices.

In addition to clinical evaluation, medical devices can also prove their effectiveness and safety through non-clinical evaluation methods. Non-clinical evaluation refers to comparing items such as the basic principles, structural composition, product manufacturing materials, performance requirements, scope of application, and usage methods of the product with already marketed medical devices to prove that the product is highly comparable to the marketed devices.

To conduct medical device clinical trials in the Macao Special Administrative Region, prior approval from the ISAF is required. According to the "*Detailed Rules for the Implementation of the Medical Devices Supervision and Administration System*" which will take effect together with Law No. 12/2025 on July 1, 2026, the following documents and materials must be submitted when applying: Applicant information and proof of legitimacy, documents proving the clinical trial institution complies with Good Clinical Practice (GCP) or equivalent standards, device research background documents and general technical data, clinical trial protocols, investigator manual, statistical analysis plan, informed consent form for subjects, relevant documents of the ethics committee, and samples of the labels and instructions.

It should be noted that for the import of medical devices, the relevant clinical evaluation materials must be submitted to the ISAF during the registration process. Only clinical trial data obtained from domestic and foreign institutions that comply with the medical device Good Clinical Practice norms, such as the GCP in China mainland and the ISO 14155 standards in Portugal and the European Union will be accepted by the ISAF. As such, applicants can choose to conduct multi-center clinical trials simultaneously in the local area and other places without having the entire clinical trial conducted in Macao.

In principle, medical devices must undergo clinical evaluation. However, for those devices with a clear working mechanism, mature design, and no serious adverse events recorded in clinical applications of the same type over many years, or those that can prove their safety and effectiveness through non-clinical methods, clinical evaluation exemption can be legally granted. Nevertheless, regarding the application or determination procedures for exemption, the current legislation lacks clear stipulations. The ISAF is expected to clearly define the application procedures, determination standards, and review mechanisms for exemption when formulating relevant technical guidelines.

5.4.4 Activity Licenses

Medical device business activity licenses are divided into medical device manufacturing licenses and medical device operational licenses. Applicants must apply for the corresponding license based on the business activities they are engaged in, rather than based on the product category. According to Law No. 12/2025, only after obtaining the manufacturing license can one engage in the manufacturing business of medical devices in the Macao Special Administrative Region.

The application procedure for a manufacturing license consists of two stages: plan review and site inspection. To apply for or renew a license, the applicant must meet the following conditions simultaneously: a natural person must have a residence in Macao or a legal person must be legally established in Macao; the site must hold an industrial license issued by the Economic and Technology Development Bureau and possess the corresponding conditions for manufacturing medical devices, including meeting the prescribed intervals, facilities, and equipment; the applicant must establish a quality management system suitable for the manufactured devices in accordance with Good Manufacturing Practices and possess the corresponding after-sales service capabilities; the site must be equipped with a qualified Technical Supervisor; in addition, the applicant must not be under an industry prohibition period and must not have debts undergoing tax enforcement procedures.

Regarding entrusted manufacturing, if the applicant, registered holder or registrant entrusts others to manufacture medical devices, prior approval from the ISAF is required. This regulation aims to ensure that the manufacturing factory (whether in Macao or outside Macao) entrusted by the registered or registrant applicant possesses the appropriate structure, personnel and quality management system that matches the entrusted medical devices and has the capability to manufacture the medical devices for which the registration or filing is applied. For such an application, the following documents must be submitted: draft of the entrusted contract, proof of the entrusted party's capabilities, and other materials required by the ISAF. It should be noted that the law prohibits the entrusted party from re-entrusting the manufactured medical devices to a third party for further manufacturing.

The operational license is an administrative license issued for a specific location, covering sales activities (wholesale or retail) and import/export activities. Law No. 12/2025 stipulates that an operational license must be obtained before engaging in the import, export, and wholesale of Class IIb and Class III medical devices, and the retail of Class IIb medical device. To simplify the application procedure and improve the administrative efficiency, the "*Detailed Rules for the Implementation of the Medical Devices Supervision and Administration System*" has established a one-stop licensing procedure for operational licenses and promoted the use of a unified electronic platform. The aim is to integrate the cross-departmental joint approval processes among authorities such as the ISAF, Labor Affairs Bureau, Fire Services Bureau, and Public Works Bureau, thereby shortening the approval time and achieving the goal of facilitating the public and businesses. The application procedure for an operational license includes two stages: plan review and site inspection. The application or renewal of an operational license is similar to that for the manufacturing license, including requirements for the applicant's local residence or registration status, absence of disqualification orders and no tax arrears. Additionally, it is necessary to ensure that the spacing, facilities, and equipment of the business site comply with regulations, and qualified Technical Supervisors must be equipped. Moreover, the business site must not be located within real estate used for residential or office purposes.

In terms of the site layout, medical device manufacturing facilities must have administrative offices and storage space for medical devices; facilities engaged in the operation of Class IIb or Class III medical devices must have dedicated areas, facilities, or equipment for storing such devices. Any future changes, such as changes to the workplace or company personnel, must be applied for permission or registered with the ISAF in advance. Regarding the validity period of the licenses, the manufacturing license is valid for three years and can be renewed for the same period. The operating license is valid from the date of issuance until the end of December of the following year and can be renewed annually thereafter. The licensing requirements for manufacturing, importing/exporting, wholesale, and retail of medical devices across different risk categories are indicated in Table 5.3.

Table 5.3 Regulatory requirements for different classes of medical devices

Category	Market Authorization	Manufacturing	Import/ Export & Wholesale	Retail
Class I	Listing	Manufacturing License	No Operational License required	No Operational License required
Class IIa	Listing	Manufacturing License	No Operational License required	No Operational License required
Class IIb	Registration	Manufacturing License	Operational License required	Operational License required
Class III	Registration	Manufacturing License	Operational License required	Retail Prohibited

The Technical Supervisor (TS) is a professional position mandated by law, responsible for providing technical guidance and quality control at medical device manufacturing sites or Class IIb medical device retail sites. Without the TS, an enterprise will not be able to obtain or maintain its business license. Depending on the nature of the business at the site, the TS is required to have corresponding educational and work experience requirements. For instance, the TS for a medical device manufacturing plant must hold a bachelor's degree or equivalent (including consecutive master's or doctoral degrees) that is compatible with the produced devices and possess at least three years of relevant work experience in medical device manufacturing and quality management. In terms of the recognition of work experience, it is particularly important to note that the three years of relevant work experience required by the TS do not include internship experiences during the degree program, nor do they cover the relevant work experience accumulated before obtaining the degree qualification. The TS for Class IIb medical device retail site must hold a relevant university degree or have at least three years of relevant work experience.

The ISAF will conduct strict and dynamic monitoring of the on-duty status of the TS. Enterprises must strictly adhere to the relevant reporting deadlines: for short-term absences, the license holder can appoint a person with equivalent qualifications to perform the duties on their behalf and must notify the ISAF in advance. If there are valid reasons, the report must be submitted no later than five days after the incident. If the TS is absent for more than 90 days or leaves the position, the enterprise must apply to the ISAF for a replacement candidate at least 15 days in advance (30 days for valid reasons). Failure to submit the report on time will directly result in the ISAF classifying the site as "without a Technical Supervisor", and the enterprise's operation license will face the risk of suspension or revocation.

5.4.5 Post-Market Evaluation

In terms of the enforcement of post-market supervision, inspectors from the ISAF are vested with the powers of a public authority. When necessary, they may request assistance from law enforcement and administrative authorities in accordance with the law to ensure the effective implementation of regulatory activities. The timely reporting of adverse events is a core component of post-market evaluation. Once a registration holder or filer becomes aware of or suspects a serious adverse event involving a medical device, they must report it to the ISAF within 24 hours. After receiving a report or obtaining relevant information on its own, the ISAF may further require the registration holder or filer to submit a detailed report within a specified period to facilitate an in-depth assessment.

The registration holder or filer must proactively conduct a post-market evaluation of the medical device under the following circumstances:

(1) Based on the development of scientific research, there is a change in the understanding of the safety and effectiveness of the medical device.
(2) The results of medical device adverse event monitoring and assessment indicate that the medical device may have defects.

5.4.6 Transitional Arrangements

The overall approach aligns with Singapore's regulatory framework for medical devices, with products being phased into the regulatory system based on risk levels (Figure 5.2). The goal is to gradually complete the market's transition to compliance while ensuring that the supply of medical devices in Macao remains unaffected. The specific transition provisions are as follows:

Medical device manufacturing sites: Existing manufacturing plants must apply for manufacturing licenses by specific deadlines based on the risk level of the products they engage in. For Class III and Class IIb medical device manufacturing businesses, the application must be completed by July 1, 2027. For Class IIa, it must be completed by July 1, 2028, and for Class I, it must be completed by July 1, 2030. One year before the application deadline is the transition period, during which the manufacturing plants can continue to operate. If new or changed manufacturing products are introduced, a new application for the license must be made.

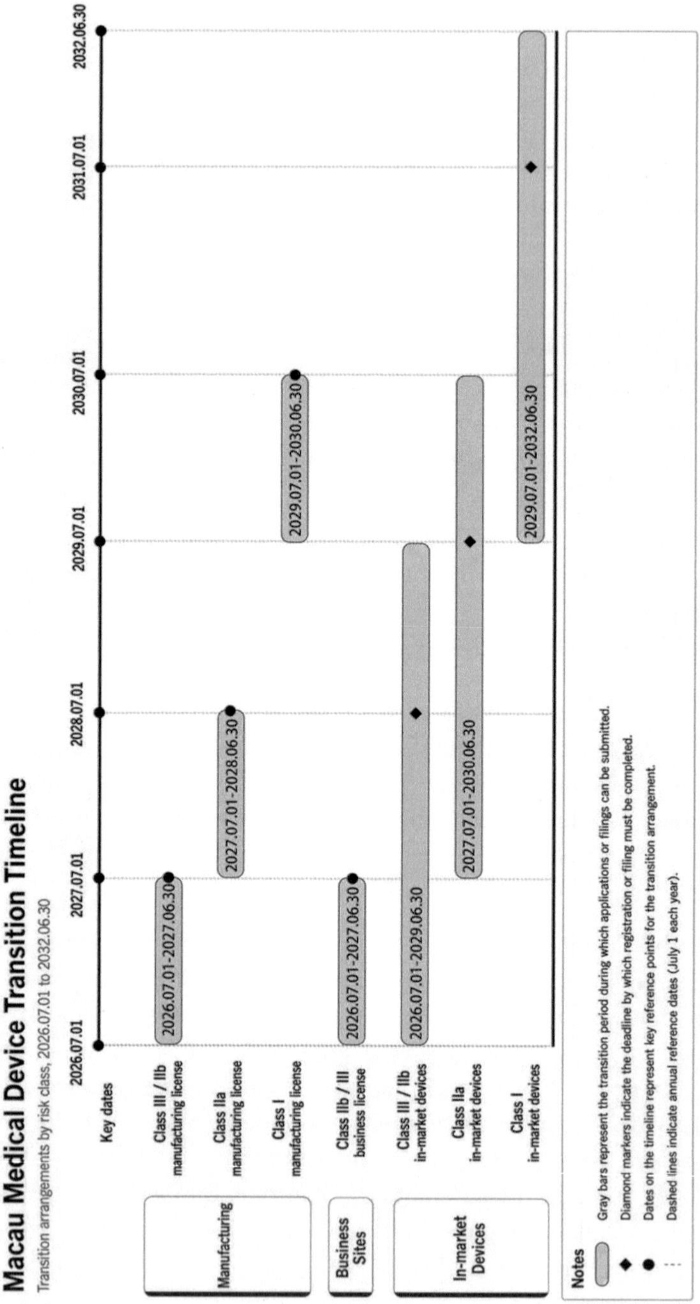

Figure 5.2 Regulatory transition framework for medical devices in Macao.

Operating premises such as medical device importer, exporter, wholesaler, or retail: Those engaged in the import, export, wholesale or retail business for Class IIb or Class III medical devices must apply for an operating license from the ISAF by July 1, 2027. Existing businesses can continue to operate during the application period as long as they apply for a license within one year after the new law takes effects on July 1, 2026. Sufficient time will be assigned to assist the industry while following the original system and without affecting the normal operation of the industry, to gradually meet the compliance requirements.

Medical Devices Already in Circulation: Transition periods are also established in batches according to risk levels: the transition period for Class III and Class IIb is from July 1, 2026, to June 30, 2029; for Class IIa, from July 1, 2027, to June 30, 2030; and for Class I, from July 1, 2029, to June 30, 2032. During the transition period, relevant devices may continue to be manufactured, imported/exported, and circulated in Macao.

Medical devices in transition status may continue to be manufactured or imported and exported, or circulated in Macao. However, during this period, if the related medical devices pose a threat to public health, the ISAF may intervene at any time according to law. For medical devices in transition status, relevant interested parties must apply for registration or conduct filing according to the provisions of the new law at least one year before the expiration of the transition period; otherwise, they may not circulate in Macao after the period expires. In addition, the *"Detailed Rules for the Implementation of the Medical Devices Supervision and Administration System"* provide that existing premises engaged in medical device business activities may continue to use their current signboards during the transition period, which helps reduce unnecessary costs and support the transition to the new regulations.

5.5 Conclusion

The new medical device regulatory system in Macao marks a milestone in its transition from fragmented administration to a science-based system. As a strategic fulcrum for integration into the Greater Bay Area, the new regulatory framework will help accelerate regional market integration through international benchmarking. For the industry, proactive compliance has transformed from a burden into a competitive asset. This reform will not only promote the moderate diversification of Macao's economy but also transform it into a core regulatory hub of the GBA. In an increasingly complex global health environment, it is envisioned that Macao will see success in finding a dynamic balance between technological advancement and public safety.

Chapter 6

Macao in Practice: Challenges and Enablers to Support the Implementation of the Medical Device Law in Macao

Junnan Shi, Hongyu Xian, and Carolina Oi Lam Ung
Centre for Pharmaceutical Regulatory Sciences, University of Macau, China
yc27504@connect.um.edu.mo, mc56240@connect.um.edu.mo, carolinaung@um.edu.mo

6.1 Introduction

The implementation of Macao's first legal framework for medical devices marks a significant shift in its regulatory paradigm from the traditional "trade-oriented" to "science-driven" approach. While this transformation opens up strategic opportunities to bolster the development of the health industry, it also presents profound systemic challenges. The long-standing regulatory vacuum void in medical devices requires stakeholders such as small-medium enterprises (SMEs) and academic institutions to adapt to mandatory registration and stringent clinical evidence requirements in the short term. Furthermore, heavy reliance on imports and the inertia

Medical Device Regulation in the Guangdong-Hong Kong-Macao Greater Bay Area: A Practical Guide for Compliant and Efficient Market Entry
Edited by Jack Wong and Raymond K. Y. Tong

ISBN 978-981-5352-38-2 (Hardcover), 978-1-042-03128-3 (eBook)
www.jennystanford.com

DOI: 10.1201/9781042031283-8

of existing supply chains increase the difficulty of compliance transition. This analysis aims to explore how systematic measures can be prioritized to address the transitional challenges during the implementation of Macao's medical device law.

6.1.1 Enhancing the Regulatory Capacity

Upon the enactment of the "*Law No. 12/2025—Medical Devices Supervision and Administration System*," the first key challenge would lie in the gap between technical review capabilities and regulatory requirements. Macao has long lacked specific legal guidance with regard to medical devices, resulting in a severe shortage of qualified reviewers. In the absence of sufficient technical review capabilities, the review process for medical devices (especially the complex or high-risks products) has been proven to fall into a predicament of "insufficient depth" and "excessive timeframes." In response to this challenge, the "*Detailed Rules for the Implementation of the Medical Devices Supervision and Administration System*" provide an initial institutional response by further clarifying the composition and operation of the expert committee established under Law No. 12/2025, thereby improving Macao's technical review capacity.

On this basis, Macao can further draw on the mature medical device review experiences of the Guangdong-Hong Kong-Macao Greater Bay Area (GBA) Center for Medical Device Evaluation and Inspection of the National Medical Products Administration (NMPA), promoting deep integration of Macao's regulatory system with the Guangdong-Hong Kong-Macao Greater Bay Area in terms of technical guidelines and review standards. Simultaneously, based on regional development needs, a medical device review training base should be established, along with a sound talent recruitment and joint training mechanism, to enhance professional capabilities through standardized courses and practical exercises. Furthermore, industry-academia-research cooperation with local universities and research institutions should be deepened to build a multi-party participatory regulatory science support system, promoting consistency and administrative efficiency in the regional review mechanism and achieving integrated development of the medical device regulatory system.

At present, the Pharmaceutical Administration Bureau of the Macao SAR Government (ISAF) is actively participating in various regional and international regulatory collaboration mechanisms, paving the way for substantive pathways for regulatory knowledge transfer and capacity building. At the regional level, the ISAF continues to engage in pharmaceutical regulatory cooperation within the Guangdong-Hong Kong-Macao Greater Bay Area (GBA) framework, in an attempt to achieve regulatory interoperability through dedicated working groups and information-sharing platforms. Crucially, the ISAF maintains regular exchanges with the Guangdong-Hong Kong-Macao Greater Bay Area Center for Medical Device Evaluation and Inspection of the NMPA, using this platform to coordinate technical review standards and testing specifications within the region. Furthermore, the ISAF is strengthening its alignment with Hong Kong's reformed drug and medical device regulatory system, exchanging in-depth experiences on risk-based management, registration pathway optimization, and post-market surveillance mechanisms.

In terms of training and academic development, the ISAF collaborates closely with the Advanced Training Institute of the NMPA to provide regulatory personnel with systematic regulatory training courses aimed at accelerating their professional competence in core areas such as clinical evaluation, product classification, and quality management systems (QMS). Meanwhile, the establishment of the Center for Pharmaceutical Regulatory Sciences at the University of Macau marks a significant step towards the institutionalization of regional regulatory sciences education and research, effectively bridging the scientific and academic advancements with regulatory sciences education. The Centre supports local talent development while also providing empirical support tailored to Macao's unique regulatory context. Internationally, the ISAF actively participates in international regulatory forums and maintains deep regional and international collaboration with the Drug Information Association (DIA) to ensure that regulatory personnel stay abreast of global developments.

To ensure the smooth implementation of the new medical device law in Macao, companies must also move swiftly to ensure an adequate talent supply chain. While Macao's existing workforce possesses rich market and trade experiences, they urgently need skills reshaping to meet the "science-driven" requirements of the new law. Companies should proactively encourage their quality assurance (QA) and pharmaceutical professionals to expand into the medical device field, strengthening their core functions in clinical evaluation and risk management. This is not only a legal requirement but also a technical threshold that distinguishes it from traditional drug management and embodies the "full lifecycle management" of medical devices. Companies should establish clear regulatory affairs (RA) career development pipelines, repositioning RA from their previous "administrative and logistical" role to a "strategic core" of the company. Competitive compensation packages and professional advancement channels should be provided to attract and retain scarce local talent. During this critical period of regulatory transition, companies may consider hiring senior consultants proficient in NMPA and Greater Bay Area review practices. These expert resources not only guide companies in establishing technical dossiers that meet international standards, but also help local teams "clear pitfalls" in practical application. Multinational corporations and large distributors should make good use of their established RA teams for efficient transfer of professional knowledge.

6.1.2 Optimizing the Transparency of Selecting the Expert Committee

Under the new medical device regulatory framework in Macao, the specialized committees and related functional departments under the Drug Regulatory Authority undertake the core functions of technical review and external consultation. Their authority covers authoritative verification of administrative access matters such as medical device registration, modification, and cancellation, and comprehensive assessment of product quality, effectiveness, and safety. The *"Detailed Rules for the Implementation of the Medical Devices Supervision and Administration System"* has clearly stipulated that the committee is composed of at least eleven odd-numbered professional members, including one representative from the ISAF, two representatives from the Health Bureau, two representatives from the NMPA, two representatives from the

Guangdong Medical Products Administration, and other experts from academic institutions, with a maximum term of three years and the possibility of renewal. It covers six major professional domains, including active medical devices, non-active medical devices, in vitro diagnostic reagents, clinical trials and biostatistics, biomedical engineering, and medical software. This arrangement represents an important step toward institutionalizing Macao's technical review mechanism. Although the committee's composition has been specified, its detailed review procedures, the expert selection system, and conflict-of-interest avoidance mechanism still require further clarification and refinement to ensure transparency and professionalism in administrative decision-making.

By drawing on the operational experiences of the advisory committees of the U.S. Food and Drug Administration (FDA), conflict of interest (COI) review faces a dilemma: overly stringent review mechanisms often create a high talent screen-out effect excluding leading experts with extensive industry backgrounds and weakening the committee's technical review capabilities; conversely, overly lenient review thresholds may induce regulatory capture risks, causing experts to exhibit academic bias in decisions regarding product access or risk withdrawal. For Macao, given its small scale and highly integrated medical, research, and industrial ecosystems, with extremely close networks of interests and interpersonal relationships, achieving a dynamic balance between "local expertise" and "procedural fairness" is a key issue in constructing Macao's medical device review system.

Besides COI, an imbalance in the selection criteria for the expert committee could be another cause of concern. If the selection criteria are too rigid, given the already scarce interdisciplinary experts, there is a risk of a "lack of talent," making it difficult to build a review team with multidimensional perspective as needed. However, lowering the selection threshold to fill seats will inevitability lead to uneven professional competence among the committee members, directly undermining the overall scientific rigor and authority of the technical assessment.

Furthermore, due to Macao's relatively homogeneous talent structure, it is difficult to build a top-tier expert team spanning multiple disciplines in the short term. When facing the review requirements for high-tech or high-risk (Class III) medical devices, the SAR government often relies on cross-border experts for technical consultation and professional intervention. However, while this cross-regional technical collaboration mechanism can compensate for the lack of local professional capabilities, it may also prolong the approval cycle due to the complexity of administrative coordination and review processes. This not only hinders the rapid market entry of innovative medical devices but may also weaken Macao's competitive advantage in the regional health industry.

To further enhance regulatory transparency and administrative efficiency, the ISAF could draw on international best practices (such as the FDA model) to refine the standardized selection and review system. This system should further clarify expert eligibility, evaluation processes, and entry criteria, adopting a "risk management" strategy rather than simply "risk exclusion" to ensure that experts with deep industry experience can contribute their professional insights. Furthermore, committee members should be required to sign a COI Disclosure, and a regular public complaint and oversight mechanism should be established. Through procedural transparency, the committee's external credibility will be strengthened, ensuring the scientific rigor and neutrality of administrative decisions.

In terms of talent and technology integration, companies should deepen their connections with regulatory science experts at universities, as their role as a core talent pool for specialized committees is crucial. Through early academic communication, companies can submit more scientifically structured and rigorously argued technical dossiers, effectively reducing the frequency of consultations during technology review, thereby shortening the approval cycle and accelerating product launches. For SMEs and startups, it is recommended to establish R&D compliance partnerships with industry incubation platforms at universities and research institutes, in order to leverage their bridging role and accelerate the commercialization of scientific and technological achievements. Furthermore, companies with FDA or EMA consulting experience are encouraged to act as active stakeholders, sharing international best practices to help ISAF optimize the regulatory ecosystem and build long-term public-private partnerships based on professional trust.

6.1.3 Leverage of the Real-World Data and Real-World Evidence Development

The application of real-world data (RWD) and real-world evidence (RWE) in medical device regulation has become an important trend in international regulatory development and an important tool in regulatory science in recent years. Compared to the mature real-world data conversion systems in international mature markets, Macao still presents a blank in this regard. Due to a long-term lack of systematic regulation, the local clinical data collection and monitoring mechanisms are not yet fully developed.

Although the new law has explicitly stipulated the need for post-market evaluation and adverse event monitoring, the corresponding clinical evaluation technical guidelines have not yet been issued. The collection standards, quality requirements of real-world data, and how to convert them into regulatory evidence still lack local experience. The Hainan Boao Lecheng International Medical Tourism Pilot Zone (Hainan Lecheng Pilot Zone) is the first RWD application pilot in China. It has already established a mature operation system. Macao can draw on the experience of the Lecheng Pilot Zone, implement pilot practices, and allow urgently needed imported medical devices to be used under special permits in Macao or used in the early stage, with the RWD generated during this process serving as supplementary evidence for the registration application in Macao.

In addition, Macao can consider establishing a unified real-world data platform and standardized data collection and governance norms, in order to facilitate the connection of data systems of local medical institutions. Through system integration, the problem of local data islands and the quality and compliance sharing of local medical data can be solved from the source, thereby promoting the conversion of RWD. Medical device manufacturers should adopt a proactive strategy, developing post-market data collection plans before waiting for official "real-world data (RWD)" guidelines. Companies should prioritize establishing structured electronic adverse event (e-AE) and clinical outcome tracking systems, ensuring their technical standards align with the NMPA and ISAF's anticipated data governance regulations. This will ensure that data accumulated during the transition period serves as supplementary evidence for registration.

Simultaneously, medical institutions (especially hospitals and community health centers) should accelerate the standardization of electronic health records (EHRs). Institutions with highly interoperable data foundations will have a head start in future RWD platform integration and will be prioritized for access to advanced devices under the "licensing" pilot program. For the access pathway of innovative devices, the "Hainan Lecheng Model" can be referenced: implementing high-quality managed access programs in Macao and collecting RWD under a scheme approved by the ISAF pre-approval will significantly shorten the time from first-in-human (FIH) trials to formal approval.

6.1.4 Deepening of Regional Collaboration

At present, there is a structural imbalance in the cross-border supervision between the mainland and Macao. The mainland side, through policies such as the *"Work Plan for Regulatory Innovative Development of Drugs and Medical Devices in the Guangdong-Hong Kong-Macao Greater Bay Area (GBA)"*, has opened up the access path of "Macao Registration + Hengqin Production", and clarified some regulatory divisions. However, Macao's Law No. 12/2025 focuses on local production or import-export trade, leaving legislative gaps in cross-border extension supervision, responsibility definition in the production process, and joint inspection mechanisms. This has led to insufficient institutional connection.

Furthermore, the existing regulatory information-sharing mechanism among Guangdong, Hong Kong, and Macao urgently needs optimization. The three regions have not yet established standardized mutual recognition mechanisms in key areas such as adverse event monitoring, risk warning, and product recalls, resulting in regulatory data remaining in "information silos" and hindering effective integration and coordination. This institutional imbalance has become a major obstacle to regulatory collaboration in the Greater Bay Area. Only through regulatory integration and institutional alignment can regional barriers be broken down, promoting the implementation of unified regulatory standards and thereby ensuring the safe and efficient operation of the medical device market in the Greater Bay Area.

To this end, Macao should consider promoting the supplementary and improvement of Law No. 12/2025, clarifying the scope of application and responsibility determination of cross-border extension regulation, and filling the regulatory gap in the manufacturing process of "Macao Registration + Hengqin Production". At the same time, the regulatory authorities of Macao and Hengqin may further consider bilateral regulatory cooperation agreements, clarify the joint inspection procedures and mutual recognition mechanism for results, and deepening the regulatory collaboration between the two regions.

Following the policy framework of the "*Administrative Regulations on the Pharmaceutical and Medical Devices of Nine Municipalities in Guangdong Province in Guangdong-Hong Kong-Macao Greater Bay Area Imported from Hong Kong and Macao*," Macao is actively integrating into the regional collaborative regulatory system. By establishing a standardized adverse event reporting mechanism and an integrated data exchange system, the aim is to break down "information silos" and achieve deep integration of regulatory resources. This not only enhances the overall regulatory efficiency of the GBA but also has profound strategic significance for promoting the smooth flow of drug and medical device resources and ensuring their quality and safety within the region.

For companies adopting the "Macao registration + Hengqin production" model, given the increasingly clear regulations regarding cross-border supervision and joint inspections, the primary task is to establish a clear framework for defining the rights and responsibilities between the Macao Registrant (MAH) and the Hengqin manufacturer. Any ambiguity in the definitions of responsibility in the production process will pose significant compliance risks. Companies should proactively benchmark their products against Macao registration standards and the adverse event reporting systems in the GBA to bridge any gaps in implementation. It is recommended that companies adopt an effective bilateral communication strategy, maintaining coordination with both the ISAF in Macao and the Guangdong Provincial Drug Administration. This not only demonstrates a high level of compliance integrity but also allows them to gain a strategic advantage whenever the regulatory mutual recognition mechanism progresses. Furthermore, Hengqin manufacturers should immediately optimize their quality management systems (including GMP compliance) to align with Macao registration requirements to avoid delays in market launch due to system incompatibility during future joint inspections.

6.1.5 Strengthening of the Active Monitoring System

Post-market surveillance (PMS) is a crucial component of the entire lifecycle management of medical devices. However, Macao's current regulatory framework still heavily relies on a "passive surveillance" model. Although the new law establishes the proactive reporting obligations of medical institutions and Marketing Authorization Holders (MAHs), this mechanism, centered on "spontaneous reporting," often faces inherent risks of underreporting, missed reporting, or delayed reporting, reflecting a gap in the regulatory system's ability to identify and proactively intervene in early risks.

The significant lag in passive surveillance means that by the time the ISAF becomes aware of a risk, damage has often already occurred. This poses a significant risk, particularly for high-risk (Class III) medical devices, potentially leading to a public health crisis. Therefore, a shift in regulatory paradigms, constructing a multidimensional proactive surveillance system based on "deep surveillance" is recommended, using innovative mechanisms to overcome the limitations of spontaneous reporting and achieve precise early warning and dynamic control of risks.

First, a comprehensive Unique Device Identifier (UDI) system could be considered to establish a precise traceability path throughout the entire lifecycle. Building upon this, the standardized interoperability of EHRs and cross-platform medical data could be promoted to achieve deep integration of multi-source data. Simultaneously, artificial intelligence algorithms such as deep learning should be introduced as a core driving force for regulatory innovation, enabling in-depth mining and risk modeling of massive amounts of clinical information, providing precise empirical support for early warning of adverse events. Ultimately, this will shape a data-driven, closed-loop response, and automatically iterative intelligent proactive monitoring system. As medical technologies evolve and clinical application experience accumulates, the regulatory system must establish a dynamic evaluation mechanism based on this system. Through continuous data feedback, regulatory strategies can be flexibly adjusted to accurately align with the pulse of industrial innovation and public health defense needs.

Industry should view the voluntary adoption of UDI systems as a strategic competitive advantage and risk management tool, rather than simply a compliance burden. A supply chain empowered by UDI can significantly improve the accuracy and response speed of adverse event investigations, thereby reducing legal liability risks and operational disruptions caused by product recalls. For this, it is important for the medical device manufacturers to convert product data into a standard format compatible with EHRs. For AI-driven medical device companies, building internal AI-assisted post-market surveillance capabilities (such as clinical outcome signal detection algorithms) is crucial. This not only meets the requirements of a proactive surveillance framework but also establishes a company-owned Real-World Evidence Database (RWE Database), providing strong data support for future registration and renewals. Furthermore, maintaining transparent and timely adverse event reporting records will help companies and regulatory agencies build high levels of mutual trust, thereby optimizing approval efficiency.

6.1.6 Empowerment of the Regional Synergies

The promulgation of the new medical device regulatory framework in Macao represents a historic leap in systems strengthening for ensuring public safety in the use of medical devices. Moreover, it also breaks the invisible constraints of traditional trade models on the development of high-tech industries, providing core institutional support for Macao's deep integration into the GBA's health industry landscape and the implementation of its economic diversification strategy. Specifically, this reform significantly strengthens regional synergy in the following three dimensions:

1. System synergy

To further implement the Outline Development Plan for the Guangdong-Hong Kong-Macao Greater Bay Area, the NMPA and other departments issued the "*Work Plan for Regulatory Innovation and Development of Pharmaceutical and Medical Device in the Guangdong-Hong Kong-Macao Greater Bay Area (GBA)*" in 2020. The core of this plan is to allow designated healthcare institutions operating in the

GBA to use medical devices for urgent clinical purposes used in the public hospitals in Hong Kong and Macao, subject to the approval of Guangdong Province (the Measure). The Measure is currently the most important cross-border policy mechanism for pharmaceuticals and medical devices in the Greater Bay Area. As of December 2025, the designated medical institutions under the Measure have reached 71. A total of 140 drugs and medical devices from Hong Kong and Macao have been approved for use in the GBA through this pathway, benefiting approximately 17,000 patients and covering nine cities in the mainland of the Greater Bay Area.

However, behind the prosperous data lies an imbalance in regional collaboration. According to the "*Clinical Urgent Need Import Catalogue of Hong Kong and Macao Medical Devices for the Nine Mainland Cities of the Guangdong-Hong Kong-Macao Greater Bay Area (2025 Edition)*," the medical devices introduced to the mainland almost all originated from Hong Kong, with very few from Macao. The introduction of the new law in Macao provides clear legal basis and quality guarantee for Macao-registered medical devices to enter the directory of the "Measure" and be used in medical institutions in the mainland of the GBA. This act has transformed Macao from a marginal participant in this cross-border mechanism to an important one.

2. Industry synergy

The "*Several Measures Supporting the High-Quality Development of Biopharmaceutical Big Health Industry in the Guangdong-Macao in Depth Cooperation Zone*" clearly proposed to build the brand "Macao registration + Hengqin Production." Macao, leveraging its status as an independent customs territory and its unique positioning as an open platform, undertakes the functions of registration and brand exporting, especially having a circulation advantage that mainland brands cannot replace in terms of reaching Portuguese-speaking countries and markets along the "Belt and Road" initiative.

Meanwhile, Hengqin relies on abundant land and generous financial subsidies to undertake large-scale production and manufacturing. The introduction of the new law in Macao fills the legal gap for "Macao Registration" in the medical device field, enabling this division of labor model to be verified from the pharmaceutical field and extended to the medical device industry, promoting the deep integration of the medical device industries in both Macao and Hengqin.

3. Industry-university-research synergy

Technological innovation is the key driving force for the development of industries. Universities, research institutions and related industrial parks serve as the main sites for innovative research. In 2019, the University of Macau established the first industry-university-research cooperation base in the Greater Bay Area in Hengqin-Zhuhai UM Science and Technology Research Institute. Relying on the research capabilities of University of Macau's three national key laboratories and advantageous academic institutions, the institute conducts applied scientific research and technology transfer work in Hengqin.

Macao's medical device manufacturers and importers/exporters should immediately assess whether their products meet the access standards of the "*GBA Catalogue of Clinically Urgently Needed Drugs and Medical Devices*." Companies should proactively engage with the ISAF and the corresponding authorities in Guangdong and Hong Kong to initiate coordination for inclusion in the catalogue. As Macao transitions from a "passive participant" to a "core collaborator" in cross-border policies, companies are facing a critical policy window. Early inclusion in the catalogue means that companies can gain priority access to designated medical institutions in nine cities in mainland China. For companies already registered in Hong Kong but not yet with a physical presence in Macao, "*Macao registration*" has now become an independent and exclusive pathway to mainland China; under the current legal framework, "dual registration in Hong Kong and Macao" possesses extremely high strategic appeal.

6.2 Conclusion

In conclusion, Law No. 12/2025 is not only a milestone in the local regulatory science development of Macao, but also a legal support for Macao's deep integration into the medical device innovation ecosystem of the Greater Bay Area. Looking ahead, driven by the continuous improvement of regulatory science and regional collaboration, Macao will surely leverage its unique institutional advantages to become a core hub connecting the mainland and the international frontier medical markets, injecting strong industrial momentum into the full implementation of the "1+4" economic moderate diversification development strategy.

Part III

Greater Bay Area: Integration and Strategy

Chapter 7

Greater Bay Area: Using Hong Kong and Macao as Gateways

Edward F. W. Yau
Member, Hong Kong Local Responsible Persons Panel
edward9988@gmail.com

7.1 Introduction: The Strategic Importance of the Greater Bay Area

The Guangdong-Hong Kong-Macao Greater Bay Area (GBA) is a monumental regional development initiative and a foundational cornerstone of China's national strategic blueprint. Comprising the Hong Kong Special Administrative Region (SAR), the Macao SAR, and nine municipalities in Guangdong Province—namely Guangzhou, Shenzhen, Zhuhai, Foshan, Huizhou, Dongguan, Zhongshan, Jiangmen, and Zhaoqing (the nine PRD municipalities)—the GBA covers an expansive total area of 56,000 km^2. At the end of 2017, the combined population of this highly integrated region stood at approximately 70 million, and its gross domestic product reached

Medical Device Regulation in the Guangdong-Hong Kong-Macao Greater Bay Area: A Practical Guide for Compliant and Efficient Market Entry
Edited by Jack Wong and Raymond K. Y. Tong

ISBN 978-981-5352-38-2 (Hardcover), 978-1-042-03128-3 (eBook)
www.jennystanford.com

DOI: 10.1201/9781042031283-10

around RMB 10 trillion, securing its status as one of the most open, economically vibrant, and technologically advanced regions in China [1].

The overarching goal of the GBA development is to build a world-class city cluster and an international first-class bay area that serves as an ideal place for living, working, and travelling. To fulfill this vision of a "quality living circle," Chinese regulatory authorities recognized the urgent need to integrate the disparate healthcare, legal, and economic systems of the Mainland, Hong Kong, and Macao. Central to this healthcare integration is the Measure of using HK-registered drugs and medical devices used in HK public hospitals in Guangdong-Hong Kong-Macao Greater Bay Area (the GBA measure). This innovative policy allows specifically designated healthcare institutions operating within the nine mainland GBA cities to import and utilize clinically urgent drugs that are currently registered in Hong Kong and Macao, alongside medical devices that are actively used in Hong Kong and Macao public hospitals [2].

To fully understand the profound impact and necessity of the GBA measure, it is essential to exhaustively examine the broader macro-level health policies of China, the regional development outlines of the GBA, the historical regulatory disparities between Hong Kong and the Mainland, Hong Kong's recent dynamic evolution into an international medical innovation hub, and the comprehensive legislative frameworks and application procedures that govern this unprecedented cross-border healthcare integration [1, 2, 8].

7.2 The Macro-Level Health Policy Context of China: Healthy China 2030

The foundation of the GBA measure is deeply rooted in China's evolving national health policies, which have fundamentally shifted the national focus from disease treatment to comprehensive population health management. The "Healthy China 2030" Planning Outline, released by the Central Committee of the Communist Party of China and the State Council in 2016, represents the highest-level strategic blueprint for the nation's health sector. The core strategic theme of this outline is "Co-construction and Sharing, Health for All". It firmly establishes health as a prerequisite for comprehensive human development and a foundational condition for economic and social progress [3].

The outline sets ambitious, quantifiable targets to be achieved by 2020 and 2030. For instance, the average life expectancy is targeted to increase from 76.34 years in 2015 to 77.3 years by 2020, and further to 79.0 years by 2030 [3]. The infant mortality rate is projected to decrease from 8.1% in 2015 to 5.0% by 2030, while the under-5 children mortality rate is set to drop from 10.7% to 6.0%. Maternal mortality is expected to fall from 20.1 per 100,000 in 2015 to 12.0 per 100,000 by 2030. Furthermore, the policy aims to elevate the national health literacy rate from 10% in 2015 to 30% by 2030, and increase the number of people regularly participating in physical exercise from 360 million to 530 million [3].

The "Healthy China 2030" outline also emphasizes the reduction of premature mortality from major chronic diseases by 30% compared to 2015 levels by the year 2030. To achieve these goals, the government mandates the integration of health considerations into all public policies, establishing a health impact assessment system to systematically evaluate the impact of economic and social development plans on health. It also strongly advocates for the expansion of the health service industry, projecting its total scale to exceed 16 trillion RMB by 2030. The GBA measure aligns perfectly with this outline by expanding access to high-quality medical services, fostering a robust health industry, and improving overall health equity [2, 3].

7.3 The 14th Five-Year National Health Plan (2021–2025)

Building upon the "Healthy China 2030" outline, the "14th Five-Year National Health Plan" was formulated to further propel the health sector forward. This comprehensive plan emphasizes the primary principle of prevention, strengthening grassroots medical capacities, and accelerating the expansion of high-quality medical resources [4].

A critical component of this plan is the reinforcement of the public health protection network, specifically enhancing the capacity for disease prevention and control, early warning mechanisms, and emergency response to major epidemics and sudden public health events. The plan sets the goal of increasing the average life expectancy by an additional 1 year compared to the 2020 baseline [4].

The 14th Five-Year Plan also addresses the full lifecycle of population health. It highlights the optimization of birth policies, including the implementation of the three-child policy and its supporting measures, enhancing maternal and child health services, and proactive strategies to address the rapidly aging population. For the elderly, the plan aims to ensure that by 2025, over 65% of individuals aged 65 and above receive standardized health management services in urban and rural communities, and over 60% of secondary and above comprehensive hospitals establish geriatric medicine departments [4].

Furthermore, the plan champions the preservation, innovation, and development of Traditional Chinese Medicine (TCM). It outlines major engineering projects for TCM revitalization, aiming to improve the service capacity of advantageous TCM specialties at the prefectural level and above. The plan also explicitly encourages the innovation and development of the pharmaceutical industry, specifically calling for the accelerated research, development, and industrialization of drugs for major diseases and clinical emergencies, as well as the promotion of high-end medical equipment manufacturing. The GBA measure serves as a direct regulatory innovation to fulfill these exact objectives by accelerating the entry of urgently needed global drugs and high-end medical devices into the mainland system [2, 4].

7.4 The 15th Five-Year Plan Recommendations and Future Outlook

Looking ahead to the "15th Five-Year Plan" (2026–2030), recent recommendations from the CPC Central Committee underscore a continued, unwavering commitment to the "Health Priority Development Strategy". The upcoming phase will focus on strengthening public health capabilities, improving disease control systems to prevent major infectious diseases, and establishing a robust mechanism that synergizes medical care, medical insurance, and pharmaceutical governance to promote hierarchical diagnosis and treatment [5].

The 15th Five-Year Plan recommendations emphasize deepening the reform of public hospitals with a public-welfare orientation, focusing on pricing, remuneration, and comprehensive supervision, while heavily strengthening the operational guarantees for county-level and grassroots medical institutions. The plan continues to focus heavily on the "one old, one young" demographic—optimizing basic elderly care services, improving urban and rural elderly care networks, and implementing early pregnancy care actions alongside capacity enhancement plans for birth defect prevention [5].

Crucially, the 15th Five-Year Plan boldly targets the development of "New Quality Productive Forces" within the health sector [5, 6]. Regional governments are heavily investing in original and disruptive technological innovations. In the GBA specifically, Guangdong Province plans to strengthen original innovation and core technology breakthroughs in neuroscience, brain-computer interfaces, embodied artificial intelligence, smart robots, biomanufacturing, and biopharmaceuticals.

An example from the Hainan Province is pioneering a "乐城研用 + 海南生产" 模式" "Lecheng Research and Use + Hainan Production" model to accelerate the translation of pharmaceutical and medical device R&D into localized production [6]. The concept of the model is:

1. Research and Use (Lecheng Pilot Zone): Leverages special policies to allow patients early access to cutting-edge, overseas-approved drugs and medical devices that have not yet been registered in the rest of China. It also utilizes real-world data (RWD) collected here to accelerate domestic regulatory approval.
2. Hainan Production (e.g., Haikou National High-Tech Zone): Once the innovative medical products gain import registration in China, their manufacturing and supply chain processes are transferred or scaled up to localized enterprises in Hainan's industrial zones.

This model would allow pharmaceutical and medical device companies to perform trials on their products in the pilot zone and fast-track regulatory approval, ultimately providing earlier access to patients [6].

The GBA measure perfectly complements these initiatives by providing a real-world clinical testing ground (via designated GBA hospitals) for advanced global medical technologies, thereby accelerating the accumulation of new productive forces in the region [2, 6].

7.5 The Greater Bay Area Development Strategy and Healthcare Integration

The integration of health systems under the GBA measure is not an isolated regulatory adjustment but a critical, foundational pillar of the broader "Outline Development Plan for the Guangdong-Hong Kong-Macao Greater Bay Area".

Strategic Positioning of the GBA Cities The GBA is designed to be a vibrant world-class city cluster, an international innovation and technology hub, an important support pillar for the Belt and Road Initiative, and a showcase for in-depth cooperation between the Mainland and Hong Kong/Macao [1]. The Outline Development Plan carefully defines the specific roles of the core cities to maximize regional synergy [1]:

- Hong Kong: Tasked to consolidate and enhance its status as an international financial, transportation, and trade center, as well as an international aviation hub. It aims to strengthen its status as a global offshore Renminbi (RMB) business hub, promote innovation and technology industries, and establish itself as the center for international legal and dispute resolution services in the Asia-Pacific region.
- Macao: Positioned to develop into a world-class tourism and leisure center and a commerce and trade cooperation service platform between China and Lusophone (Portuguese-speaking) countries. It is also designed to be a base for exchange and cooperation where Chinese culture is the mainstream and diverse cultures coexist.
- Guangzhou: Expected to fully leverage its leading function as a national core city and an integrated gateway city, comprehensively strengthening its functions as an

international commerce and industry center and integrated transport hub.

- Shenzhen: Positioned to leverage its leading role as a special economic zone and national innovation city, striving to become a capital of innovation and creativity with global influence [1].

Shaping a "Healthy Bay Area" Chapter Eight of the Outline Development Plan explicitly focuses on developing a "Quality Living Circle for Living, Working and Travelling," featuring a dedicated section entirely focused on "Shaping a Healthy Bay Area". This section dictates the necessity of close medical and healthcare cooperation across the borders. It actively supports medical and healthcare service providers from Hong Kong and Macao in setting up healthcare facilities—through sole proprietorship, joint venture, or cooperation—in the nine PRD municipalities, thereby developing a regional healthcare cluster [1].

The strategy emphasizes the internationalization of Traditional Chinese Medicine. It supports Macao and Hong Kong in leveraging their State Key Laboratories of Quality Research in Chinese Medicines to establish internationally recognized reference standards, propelling Chinese medicine products into global markets [1].

The plan also calls for enhancing the joint nurturing of medical talents, encouraging medical professionals from Hong Kong and Macao to visit the nine PRD municipalities for academic exchanges and short-term private practice. Furthermore, the plan mandates the exploration of cross-boundary transfer services for non-emergency patients via land routes, and the establishment of a unified emergency medical rescue linkage mechanism [1]. In addition, it emphasizes cooperation in food safety, aiming to improve the traceability system for the origins of food among Hong Kong, Macao, and the Mainland, ensuring that food supplied to Hong Kong and Macao meets stringent safety standards. The GBA measure acts as the central regulatory catalyst to actualize this "Healthy Bay Area" by removing the strict regulatory bottlenecks that previously prevented the fluid movement of advanced medical therapies across the borders [2, 9].

7.6 The Pre-GBA Measure Regulatory Landscape: Hong Kong's Distinct Framework

To fully appreciate the necessity and immense legal complexity of the GBA measure, one must comprehensively understand the distinct regulatory environment of Hong Kong [7, 8]. Operating under the "One Country, Two Systems" framework, Hong Kong functions as a separate customs territory with its own deeply localized legal ordinances [1]. While pharmaceuticals face strict, well-defined import and export controls under the Pharmacy and Poisons Ordinance (Cap. 138) and the Import and Export Ordinance (Cap. 60), the landscape for medical devices is remarkably different [7].

Pharmaceutical Import and Export Controls in Hong Kong: In Hong Kong, the import and export of pharmaceutical products, medicines, and dangerous drugs are strictly governed by the Import and Export Ordinance (Cap. 60) and the Pharmacy and Poisons Ordinance (Cap. 138).

Under Section 2 of the Pharmacy and Poisons Ordinance (Cap. 138), a "pharmaceutical product" or "medicine" is broadly defined as a substance or combination of substances presented as having properties for treating or preventing disease in human beings or animals, or used to restore, correct, or modify physiological functions. To import or export these products, traders must first secure the appropriate trader licenses, such as a Wholesale Dealer Licence, a Licence for Manufacturer, or an Antibiotics Permit.

For every single shipment of pharmaceutical products, an explicit Import Licence (Form 3, typically blue) or Export Licence (Form 6, typically white) must be obtained from the Drug Evaluation and Import/Export Control Division of the Department of Health. These licenses have strict validities: 6 months for an Import Licence and 28 days for an Export Licence [7].

Hong Kong operates a robust online system for these applications, known as the Pharmaceuticals License Application and Movement Monitoring System (PLAMMS). Usually, imported pharmaceutical products must be registered in Hong Kong before they can be sold for local consumption. However, there are notable exceptions. Unregistered drugs can be imported for re-export, for manufacturing purposes, for clinical trials, or—crucially—for the treatment of a named patient. The "Named Patient Import" pathway requires a letter from a registered medical doctor stating the patient's details and required quantity, a Certificate of Analysis, product information, and proof of overseas registration.

Dangerous Drugs and Psychotropic Substances: The control of dangerous drugs (DD) is governed by the Dangerous Drugs Ordinance (Cap. 134), which aligns with international treaties such as the 1961 Single Convention on Narcotic Drugs and the 1971 Convention on Psychotropic Substances. Dangerous drugs include substances specified in Part I of the First Schedule. Certain psychotropic substances (e.g., Pentobarbital, Zolpidem) require handling by registered pharmacists.

The process for importing dangerous drugs is exceptionally stringent. The importer must first apply for an Import Certificate (a grey, three-part document). The original of this certificate is sent to the overseas supplier, who presents it to their competent authority to secure an Export Authorization. Only when the consignment is about to arrive in Hong Kong does the importer apply for the actual Import Licence (pink), which must match the quantities authorized in the Import Certificate.

Hong Kong also heavily regulates dangerous drugs in transit. Goods passing through Hong Kong to a third country must be accompanied by valid export authorizations and require a "Removal Licence" to be moved from the arriving vessel/aircraft. If a transit shipment needs to change its destination country, a highly scrutinized "Diversion Licence" is required, necessitating proof of a valid import certificate from the new destination country.

Medical Device Regulations in Hong Kong: Unlike pharmaceuticals, Hong Kong currently lacks a specific, overarching statutory legislation that comprehensively regulates the manufacture, import, distribution, supply, and use of all medical devices. Instead, specific products might fall under overlapping laws like the Pharmacy and Poisons Ordinance (Cap. 138) for devices containing medicine, the Radiation Ordinance (Cap. 303) for emitting devices, or the Telecommunications Ordinance (Cap. 106) [7, 8].

To manage this legislative gap, the Department of Health operates a voluntary Medical Device Administrative Control System (MDACS) [8]. Under this system, importers can apply to be included on the "List of Importers" using Application Form MDIP+D via the online

Medical Device Information System (MDIS). This listing is entirely voluntary. An importer is defined as a legal person who brings medical devices into Hong Kong for supply, excluding persons who bring them for personal use or those employed merely to carry the products [7].

To be listed, importers must maintain a properly manned office in Hong Kong for business operations and submit documented procedures to the Medical Device Division (MDD). These procedures must comprehensively cover ensuring the standard of imported devices, keeping supply records for traceability, proper handling, storage, and delivery, managing safety alerts, field safety notices, recalls, reporting adverse events, handling complaints, and tracking specific high-risk devices. The importer must provide the MDD with access to supply records and premises for inspection. Furthermore, any public advertising must include disclaimers stating that the importer's listing does not imply the listing of the actual medical devices or their compliance with MDACS requirements. Importers must also submit a comprehensive list of their imported devices, detailing the manufacturer, model, description, and listing number, and must notify the MDD within four weeks of any changes to this information [7].

The Dilemma: Prior to the GBA measure, this distinct, internationally aligned yet voluntary regulatory environment in Hong Kong meant that while Hong Kong patients had rapid access to cutting-edge global medical devices, patients just across the border in the mainland GBA cities did not. Mainland hospitals had to wait years for these products to pass the lengthy clinical trial and registration processes of the National Medical Products Administration (NMPA) [2, 8]. The GBA measure was conceived specifically to dismantle this regulatory barrier, creating a bridge for these advanced products to reach mainland patients seamlessly [2, 9].

7.7 Hong Kong's Strategic Evolution: Developing the International Health and Medical Innovation Hub

While the GBA measure revolutionized the flow of medical products from Hong Kong to the Mainland, Hong Kong itself is undergoing a massive, historic regulatory overhaul. Actively aligning with the state's 14th and 15th Five-Year Plans, the Hong Kong SAR government has initiated a comprehensive reform of its drug and medical device approval systems. The strategic objective is to transform Hong Kong into an authoritative "International Health and Medical Innovation Hub" (國際醫療創新樞紐), creating an ecosystem that not only accelerates local access to advanced therapies but also serves as a premium gateway to the vast mainland market via the GBA measure.

The Shift to the Primary Evaluation Approval System: Historically, Hong Kong operated under a "Secondary Evaluation" (第二層審批) system for new pharmaceutical products. Under the Pharmacy and Poisons Ordinance (Cap. 138), a new drug application required the applicant to provide registration approvals from at least two recognized reference regulatory authorities overseas [8]. This passive reliance meant that Hong Kong could not independently approve highly innovative global drugs that had not yet secured multiple overseas registrations.

To overcome this, the SAR government announced a phased transition toward a "Primary Evaluation" (第一層審批) mechanism—where Hong Kong independently evaluates and approves clinical data without relying on prior approvals from other jurisdictions [8]. As an interim stepping stone, the "1+" mechanism was implemented on November 1, 2023 [8]. Under "1+", a new drug treating life-threatening or severely debilitating diseases needs only *one* reference regulatory approval (the "1"), provided it is supported by local clinical data and recognized by local experts (the "+"). This mechanism was highly successful; by March 2026, 19 new drugs addressing unmet medical needs had been approved via the "1+" pathway, significantly reducing patient financial burdens and accelerating access to advanced oncology and hematology therapies. Notably, nearly half of these (eight drugs) were developed and first registered in the Mainland, utilizing Hong Kong's system to validate their products for broader international markets.

Building on this success, Hong Kong officially launched the fully independent "First-Tier" new drug registration mechanism on March 31, 2026 [8]. The first phase covers the extended applications of registered chemical element products (e.g., new indications, new dosages). Subsequent phases will expand to biological products and advanced therapy products, with the ultimate goal of achieving comprehensive primary evaluation across all categories by 2030.

Establishment of the Hong Kong Centre for Medical Products Regulation (CMPR): To execute this monumental shift in regulatory independence, the SAR government is establishing an overarching, authoritative regulatory body: the Hong Kong Centre for Medical Products Regulation（香港藥物及醫療器械監督管理中心, or CMPR). Following the launch of its preparatory office in June 2024, the CMPR will be formally established within 2026, initially operating under the Department of Health.

The CMPR represents a massive structural consolidation. Historically, the regulation of Western medicines (Cap. 138), Traditional Chinese Medicines (Cap. 549), and the voluntary medical device system were highly fragmented across different boards and councils (e.g., the Pharmacy and Poisons Board, the Chinese Medicine Council of Hong Kong). Legislative amendments proposed in late 2026 will transfer these sweeping regulatory powers directly to the Director of Health, executed by the CMPR. This unified, three-in-one regulatory architecture aims to eliminate administrative overlaps, streamline the decision-making chain, and build an internationally recognized, top-tier regulatory environment that drives the local R&D industry [8].

Boosting Clinical Trial Infrastructure and Real-World Studies: Recognizing that robust clinical data is the lifeblood of primary evaluation, Hong Kong has heavily invested in its clinical trial infrastructure. On November 21, 2024, the government established the "GBA International Clinical Trial Institute" (粵港澳大灣區國際臨床試驗所) within the Hong Kong Park of the Hetao Shenzhen-Hong Kong Science and Technology Innovation Co-operation Zone. This institute coordinates clinical trial resources across Hong Kong's public and private sectors. Furthermore, an advanced "Real-World Study and Application Centre" was launched in December 2025 to harness Hong Kong's high-quality medical databases for clinical research. By synergizing with the Shenzhen Park's clinical center, this "one institute, one center" model creates an integrated, one-stop clinical trial platform for global biopharmaceutical enterprises.

International Integration and ICH Membership: A critical pillar of Hong Kong's strategy is achieving global recognition. In October

2023, with the strong support of the NMPA, Hong Kong officially became an observer of the International Council for Harmonisation of Technical Requirements for Pharmaceuticals for Human Use (ICH).

To prepare for its application to become a full ICH regulatory member by Q2 2027, the CMPR is systematically overhauling its standards. By the first half of 2026, Hong Kong will implement core ICH guidelines on Pharmaceutical Stability, Good Manufacturing Practice (GMP), and Good Clinical Practice (GCP). Additionally, Good Pharmacovigilance Practices (GVP) and Good Distribution Practice standards are being rolled out and mandated across the supply chain, ensuring that Hong Kong's drug safety, quality, and efficacy benchmarks are indisputably world-class.

Synergy with the GBA Measure ("Super Connector" Role): The establishment of the CMPR and the transition to primary evaluation profoundly amplify the impact of the GBA measure. As an independent regulatory authority recognized globally, Hong Kong acts as the ultimate "super connector" (超級連繫人). For global pharmaceutical companies, securing primary approval in Hong Kong is no longer just about accessing the local market; it provides immediate eligibility to utilize the GBA measure (港澳药械通). Innovative drugs and devices approved by the CMPR can be seamlessly channeled into the 71 designated hospitals across the nine mainland GBA cities to serve millions of patients. Conversely, for mainland biotech innovators, navigating Hong Kong's ICH-aligned primary evaluation provides an internationally recognized credential, serving as a powerful springboard to export their innovations to global markets [8].

7.8 The Genesis and Evolution of the GBA Measure

The policy framework to bridge the healthcare gap between Hong Kong and the mainland officially commenced on September 29, 2020, when the State Administration for Market Regulation, the NMPA, the National Development and Reform Commission, the Ministry of Commerce, the National Health Commission, the General Administration of Customs, the Hong Kong and Macao Affairs Office of the State Council, and the National Administration of Traditional

Chinese Medicine jointly issued the "Work Plan for Regulatory Innovation and Development of Pharmaceutical and Medical Device in the Guangdong-Hong Kong-Macao Greater Bay Area" (《粤港澳大湾区药品医疗器械监管创新发展工作方案》, Document No. 国市监药[2020]159号).

This foundational document authorized the Guangdong Provincial Government to approve the use of clinically urgent, Hong Kong/Macao-listed drugs and medical devices in designated mainland GBA institutions. The core innovation was shifting the approval authority for importing these specific urgent drugs from the National Medical Products Administration directly to the Guangdong Provincial People's Government. Furthermore, the Work Plan temporarily suspended the implementation of Article 11, Paragraph 2 of the "Medical Device Supervision and Administration Regulation" within the nine GBA cities. This suspended clause previously required foreign manufacturers of Class II and Class III medical devices to submit registration application materials and origin-country marketing approval directly to the State Council's food and drug regulatory department. By suspending this, Guangdong was empowered to approve the importation of advanced devices already used in HK/Macao public hospitals [9].

The 2020 Work Plan established a phased, pilot-first approach. It initiated the program using the University of Hong Kong-Shenzhen Hospital (HKU-SZH) as the exclusive primary pilot site [9]. Following the success of the pilot, in 2021, the Guangdong Provincial MPA and the Health Commission introduced the "Interim Regulations on the Administration of Imported Pharmaceuticals and Medical Devices for Urgent Clinical Needs in the GBA" [2].

By January 2024, a total of 28 drugs and 28 medical devices had been approved for utilization in 19 designated healthcare institutions under this Measure [2]. By June 2024, the program had achieved remarkable success: 19 designated medical institutions had been approved, 67 urgent medical products had been imported, and over 7,000 patients had directly benefited from the policy. However, practical implementation exposed several operational bottlenecks, including unclear delineations of responsibilities among different government departments, coordination difficulties, and cumbersome, lengthy approval workflows. Consequently, the

Provincial Government moved to elevate the policy from an interim measure into a formal, binding legal statute [7].

7.9 The 2024 Legislative Framework: The GBA Medical Product Management Regulation

To solidify the legal backing of the GBA measure, the Guangdong Provincial Standing Committee officially passed the "Regulations on Importing Hong Kong and Macao Pharmaceuticals and Medical Devices in the Nine Mainland Cities of the Greater Bay Area" (《广东省粤港澳大湾区内地九市进口港澳药品医疗器械管理条例》) on July 31, 2024. Coming into full legal effect on December 1, 2024, these regulations systematically address previous regulatory challenges by establishing a comprehensive framework for the application, procurement, importation, distribution, and supervision of urgent medical products.

Scope and Basic Principles: Article 2 explicitly states that the regulation applies to all activities surrounding the application, procurement, importation, distribution, usage, and supervision of clinically urgent Hong Kong and Macao pharmaceuticals and medical devices within the nine mainland cities of the Greater Bay Area (Guangzhou, Shenzhen, Zhuhai, Foshan, Huizhou, Dongguan, Zhongshan, Jiangmen, and Zhaoqing) . The policy strictly targets drugs that are clinically urgent and already marketed in Hong Kong or Macao, as well as medical devices that are clinically urgent, actively procured and used by public hospitals in Hong Kong or Macao, and feature advanced clinical applications.

Governance and Coordination Mechanisms: The Regulation establishes a multi-layered, collaborative governance structure. Under Article 3, the Provincial and Municipal People's Governments are tasked with leading the supervision, coordinating the work, and handling emergencies through a unified information-sharing mechanism. Article 5 dictates that the Provincial Government must drive collaboration frameworks directly with the Hong Kong and Macao SAR governments.

Article 4 delineates departmental responsibilities: The Guangdong Provincial and the nine municipal Medical Products

Administrations (MPAs) oversee the regulatory supervision of the urgent products. The Provincial and municipal Health Commissions regulate the clinical usage and medical behaviors associated with the imported products within the hospitals. The Medical Security Bureau oversees the pricing of medical service items related to the imported products. Furthermore, Development and Reform, Finance, Commerce, Market Supervision, and Customs departments are integrated to optimize clearance procedures.

To ensure scientific rigor, Article 8 mandates the establishment of an expert review database comprising scholars and practitioners from universities, research institutes, and medical institutions to technically evaluate the clinical urgency, advanced nature, and potential usage risks of the proposed products [10].

7.10 Strict Criteria for Institutional Participation

The GBA measure does not grant a blanket authorization to all hospitals; the privilege to import and administer these advanced therapeutics is strictly limited to heavily vetted entities [10].

Criteria for Designated Medical Institutions: Under Article 6, a hospital must apply to its municipal Health Commission and be verified by the Provincial Health Commission to become a "Designated Medical Institution" [10]. The statutory criteria mandate that the institution must:

1. Be established by Hong Kong or Macao medical and health service providers via sole proprietorship, joint venture, or cooperation (National specific hospitals may also apply under special provisions).
2. Be a legally operating Level II (Secondary) or above medical institution with a valid practice license.
3. Possess specialized clinical departments and expert technical teams appropriate for the specific medical products being requested.
4. Maintain adequate storage and maintenance facilities that strictly comply with the characteristics and instructional manuals of the urgent products.

5. Establish a dedicated adverse reaction monitoring agency equipped with professionally trained, full-time personnel capable of executing pharmacovigilance responsibilities.
6. Formulate robust emergency plans and possess the capability to handle severe adverse reactions and medical accidents.

The expansion of designated hospitals has occurred in structured batches [7]:

- The First Batch (5 hospitals): The University of Hong Kong-Shenzhen Hospital (HKU-SZH), Modern Hospital Guangzhou, Guangzhou United Family Hospital, C-MER (Zhuhai) Dennis Lam Eye Hospital, and Zhongshan Chenxinghai Hospital.
- The Second Batch (14 hospitals): Included major institutions such as the First Affiliated Hospital of Sun Yat-sen University, Sun Yat-sen Memorial Hospital of Sun Yat-sen University, Nanfang Hospital of Southern Medical University, Guangdong Provincial People's Hospital, Guangzhou First People's Hospital Nansha Hospital, Guangdong Clifford Hospital, Guangzhou C-MER Dennis Lam Eye Hospital, Shenzhen Frontier Qianhai Shekou Free Trade Zone Hospital, Shenzhen Hezheng Hospital, Shenzhen C-MER Dennis Lam Eye Hospital, Zhuhai People's Hospital (Hengqin Campus), Foshan Fosun Chancheng Hospital, Dongguan Songshan Lake Tungwah Hospital, and Dongguan Guangming Eye Hospital.
- The Third Batch (25 hospitals): Expanded massively to include The First Affiliated Hospital of Sun Yat-sen University (Nansha), The Third Affiliated Hospital of Sun Yat-sen University, The Sixth Affiliated Hospital of Sun Yat-sen University, Zhongshan Ophthalmic Center of Sun Yat-sen University, Sun Yat-sen University Cancer Center, The First Affiliated Hospital of Jinan University, Guangdong Women and Children Hospital, Guangdong Provincial Hospital of Chinese Medicine, The First Affiliated Hospital of Guangzhou Medical University, Guangzhou Women and Children's Medical Center, Qianhai Life Insurance Guangzhou General Hospital, Shenzhen Hospital of Southern Medical University, Shenzhen Luohu District People's Hospital, Shenzhen Baoan District Maternal and Child Health Hospital, Shenzhen New Frontier United Family Hospital, Shenzhen Hengsheng Hospital, Peking

University Shenzhen Hospital, The Fifth Affiliated Hospital of Sun Yat-sen University, Foshan C-MER Dennis Lam Eye Hospital, Huizhou Third People's Hospital, Huizhou C-MER Dennis Lam Eye Hospital, Dongguan Tungwah Hospital, Zhongshan City People's Hospital, Jiangmen Central Hospital, and Zhaoqing Zhengda Health Rehabilitation Hospital.

Criteria for Commissioned Operating Enterprises (Business Enterprises) Under Article 11, designated hospitals are prohibited from directly importing products; they must delegate procurement, importation, and distribution to qualified Mainland-based pharmaceutical and medical device operating enterprises [10]. These business enterprises must [2]:

1. Be established in the mainland.
2. Meet the requirements of Good Manufacturing Practices (GMP) or Good Supply Practices for Pharmaceuticals or Medical Devices, and hold valid business licenses and operating permits corresponding to the scope of the imported products.
3. Demonstrate modern logistics capabilities, including warehouses equipped with special storage areas suited for the strict conditions of the products.
4. Utilize a computerized information management system for strict quality control.
5. Possess a distribution network that covers the designated medical institutions across the GBA mainland cities.
6. Hold valid import and export business qualifications and be registered with customs as a foreign trade operator.
7. Maintain a clean legal record; they cannot be listed as persons subject to enforcement for breach of trust, be involved in major tax violation cases, or have records of serious illegal acts in government procurement within the past three years.

7.11 The Pre-Approval Database System

The operational workflow for the GBA measure begins with the Pre-Approval Database. Before any hospital can request a specific product, the product must be evaluated and listed [2].

Step 1: Listing in the Pre-Approval Database: The overseas Marketing Authorization Holder (MAH) or their authorized mainland representative must submit comprehensive documentation to the online Guangdong Government Service Network portal. The Guangdong MPA reviews the submission, and if satisfied, the product is successfully listed. For drugs, the required 13-part dossier includes [2, 7]:

1. Application Form for Clinical Urgent Imported Hong Kong and Macao Drug Pre-Review Database.
2. Approval Documentation for Market Listing in Hong Kong or Macao.
3. Samples of Drug Packaging, Labels, Instructions (notarized simplified Chinese), and Complete Product Images.
4. Overview of Adverse Drug Reactions.
5. Summary of Clinical Use.
6. Risk Assessment of Racial Differences in Use.
7. Regular Benefit-Risk Assessment.
8. Clinical Urgency and Safety Assessment of the Drug.
9. Academic Guidelines/Treatment Protocols (if applicable).
10. Relevant Documents for Pharmacovigilance Procedures.
11. Introduction to Quality System.
12. Authorization Letter (from the MAH if the application is submitted by an authorized mainland representative).
13. Commitment Letter.

A similarly rigorous 12-part dossier is required for medical devices, with the crucial addition of documentation proving the device is procured and used in HK/Macao public hospitals and an evaluation proving its advanced clinical technology compared to domestic alternatives. Once approved, the product is visible to designated hospitals [2].

7.12 Detailed Application Process for Importing Drugs

Once a drug is in the database, a designated hospital can initiate Step 2: the Importation Request [2]. The operational mechanics are defined in the "Guangdong Province Greater Bay Area Nine Mainland Cities Urgent Hong Kong and Macao Drug Application Guide" (《广东省粤港澳大湾区内地九市进口港澳药品申报指南》). The process is bifurcated based on whether the drug is already listed in the official "Urgent Hong Kong and Macao Drug Catalog" (Directory) or is a new request (Non-Directory) [11].

Application for Directory (Catalog) Drugs: For drugs already on the approved catalog, the hospital must submit:

1. Application Forms: The "Basic Information Form" (Annex 1-1) detailing the generic name (INN), trade name, dosage form, specifications, package sizes, proposed quantity, port of entry, and manufacturer. The "Basic Situation Form" (Annex 1-2) outlining global marketing status, academic guidelines, target patient population, domestic existing alternatives, and the hospital's self-evaluation report.
2. Qualification Proofs: HK/Macao official marketing approvals and a statement from the MAH confirming the product's ongoing validity.
3. Institutional Specifications: Clinical guidelines, alternative treatments, usage plan explanations, departmental management processes, authorized prescriber lists, risk mitigation plans, and patient follow-up protocols.
4. Ethical Review: Patient indications, ethics committee approval, and informed consent templates.
5. Commitment Letter: A legally binding declaration (Annex 1-3) by the hospital's legal representative guaranteeing authenticity, assuming full legal liability for adverse events, and confirming the drug will only be used internally.
6. Summary Table: For repeat imports of the exact same drug, a simplified "Summary Table" (Annex 1-4) tracking past quantities, cumulative patients treated, remaining stock, and adverse reaction monitoring results is sufficient.

Application for Non-Directory (Non-Catalog) Drugs: For unlisted drugs, the hospital must provide the standard forms (Annex 1-1, 1-2, 1-3) and product samples, plus highly rigorous clinical evaluations [11].

- Operating Enterprise Proofs: The commissioned distributor's licenses and the delegation contract.
- Clinical Urgency and Safety Assessment: The hospital must systematically argue that the drug is "clinically necessary and urgent". They must prove domestic substitutes are unavailable or inferior, highlight significant patient benefits, and provide comprehensive safety risk assessments based on overseas adverse event data.

Drug Application Workflow and Timelines: The hospital submits the application via the Guangdong Smart Food and Drug Administration system. Following a format review (with a strict 5-day window for applicants to correct deficiencies), the substantive review begins [11].

- *For Directory Drugs:* The Provincial Health Commission audits the hospital's capacity and clinical urgency within five working days. The Guangdong MPA then issues final approval within another five working days.
- *For Non-Directory Drugs:* The Provincial Health Commission has 10 working days to issue its audit. Concurrently, the Guangdong MPA convenes a specialized expert panel for a 15-working-day review (which does not count towards the administrative time limit). The MPA then has 10 working days to issue final approval and synchronize the drug into the official catalog. Approved applications receive a 1-year valid import permit. Approved drugs have included advanced oncology and hematology therapies like Azacitidine, Tremelimumab-actl, Lurbinectedin, and Ropeginterferon alfa-2b [7].

7.13 Detailed Application Process for Importing Medical Devices

The regulatory mechanics for importing medical devices, governed by the "Guangdong Province Greater Bay Area Nine Mainland Cities Urgent Hong Kong and Macao Medical Device Application Guide" (《广东省粤港澳大湾区内地九市进口港澳医疗器械申报指南》), mirror the pharmaceutical framework but incorporate strictly unique criteria.

Distinct Core Criteria for Devices: Unlike drugs, which merely need marketing authorization in HK/Macao, medical devices must strictly meet three fundamental pillars under the GBA measure [12]:

1. They must be clinically urgent.
2. They must possess advanced clinical applications.
3. Crucially, the device must have already been procured and actively used by *public hospitals* in Hong Kong or Macao.

Device Application Process (Directory and Non-Directory): The hospital submits applications for either Directory or Non-Directory devices using Annex 2-1 (Basic Information), Annex 2-2 (Basic Situation), Annex 2-3 (Commitment Letter), and Annex 2-4 (Summary Table for repeat imports) [12].

For Non-Directory Devices, the evidentiary burden is immense:

- HK/Macao Public Hospital Procurement Proof: The hospital must provide official documentation explicitly proving that an HK/Macao public institution has procured and used the exact product, alongside the manufacturer's statement of valid market status.
- Detailed Technical Descriptions: Beyond packaging and labels, passive devices and In Vitro Diagnostics (IVDs) require detailed descriptions of product structure, working mechanisms, intended use, and main raw materials. Active devices require detailed explanations of key components and software functions.
- Explanation of Advanced Clinical Application: This is the critical differentiator. The hospital must submit a comparative analytical report against domestic equivalents. They must definitively prove the requested device aligns with leading international standards and is demonstrably superior to approved domestic devices in terms of basic principles, structural design, manufacturing materials, production processes, performance metrics, or overall clinical application effects.

The workflow timelines mirror those of drugs: 5 days for HC and 5 days for MPA for Directory devices; 10 days for HC, 15 days for expert review, and 10 days for MPA for Non-Directory devices [12]. If the device is classified as large medical equipment, a configuration license and a commerce department import audit are also required [10]. Approved devices have included Cardiac Resynchronization Therapy Defibrillators (G447), the EPack cooling system, Echopulse, and the TriClip G4 Delivery System [7].

7.14 Post-Approval Operations: Traceability and Risk Management

Securing approval is only the initial milestone. The GBA Medical Product Management Regulation mandates extremely rigorous post-approval compliance to ensure patient safety and supply chain integrity.

Procurement and Customs Clearance: Under Article 12, designated hospitals and their commissioned enterprises must purchase urgent products directly from overseas MAHs, medical device registrants, or their officially authorized distributors in Hong Kong and Macao to guarantee identical quality. Upon arrival at the border, Article 13 mandates that routine port-of-entry inspections by the MPA are waived for these urgent products, significantly accelerating customs clearance [10].

Traceability, UDI System, and Record Archiving: To prevent diversion, Article 19 dictates that Guangdong Province operates a comprehensive "GBA Urgent HK/Macao Medical Product Regulatory Information Platform". Hospitals and distributors must integrate their local systems with this platform in real-time. A strict "Unique Identifier" (UDI) system ensures every unit is fully traceable from importation to patient administration, achieving a closed-loop system where risks are controllable and accountability is enforceable [10]. Furthermore, transaction notes, purchase records, and inspection data must be archived for at least 2 years post-expiration. If the product has no expiration date, records must be kept for a minimum of 5 years; records for implantable medical devices must be archived permanently [2].

Informed Consent, Adverse Events, and Recalls: Under Article 15, imported products must be strictly confined to the designated hospital. Before administration, medical staff must thoroughly educate the patient about domestic alternatives and specific risks, culminating in a signed informed consent form [10]. Designated hospitals bear primary responsibility for safety (Article 16) [10]. In the event of severe adverse reactions, they must immediately trigger emergency response plans, halt usage, and report to the Health Commission and MPA. If a product is recalled overseas, Article 18 mandates that the hospital and distributor must actively coordinate a local recall, halt distribution, and report to regulatory bodies [10].

7.15 The Financial Ecosystem: Pricing and Commercial Medical Insurance

A significant challenge of the GBA measure is the financial burden on patients. Because these imported innovations bypass standard NMPA registration, they are typically excluded from the basic national medical insurance scheme. To address this, the Regulation introduces supportive pricing and insurance frameworks.

Under Article 21, non-profit designated medical institutions are granted the autonomy to establish new medical service prices associated with the administration of these urgent products. They must adhere to principles of fairness and legality and file these self-determined prices with the Provincial Medical Security Department. To further alleviate patient costs, Article 22 strongly encourages the development of commercial health insurance products. Insurance providers are urged to incorporate GBA urgent products into "inclusive commercial health insurance" schemes. Crucially, the regulations allow patients to utilize the balance in their basic medical insurance personal accounts to purchase these supplementary commercial policies, expanding the financial safety net. Hospitals are also encouraged to purchase medical liability insurance [10].

7.16 Supervision, Legal Liabilities, and Penalties

The regulatory framework is fortified by strict enforcement mechanisms. Under Article 25, the municipal MPAs conduct continuous supervision, exploring tiered regulatory models, and hold the authority to conduct extended "flight inspections" into commissioned enterprises to verify logistics and quality standards.

If a product poses an imminent threat, the Provincial MPA can instantly suspend its importation (Article 26).

The authorities employ an escalation of warnings, administrative interviews, and formal orders for correction (Article 27) against entities failing to maintain qualification standards, traceability systems, or clinical safeguards. Crucially, under Article 28, a designated hospital faces immediate removal from the approved roster if it submits fraudulent application materials, causes a major medical accident due to regulatory violations, or refuses to correct compliance failures. Any criminal acts will be aggressively prosecuted, with full civil liability enforced for damages (Article 29) [10].

Here is the draft for the additional page to be integrated into your book chapter, written in a professional academic style that matches your existing text, utilizing the concrete clinical success stories provided in the uploaded documents.

7.17 Clinical Success Stories and Therapeutic Impact of the GBA Measure

The practical execution of the "Work Plan for Regulatory Innovation and Development of Pharmaceutical and Medical Device in the GBA" (the "港澳藥械通" or GBA measure) has delivered profound clinical benefits, transforming the regional healthcare landscape by directly addressing critical unmet medical needs. By establishing a streamlined bridge across the regulatory border, the policy has successfully accelerated patient access to life-saving therapeutics, shifting clinical outcomes from protracted waiting periods to immediate, advanced interventions.

Breakthrough Oncology Drug Interventions: Folate Receptor-Targeting ADC

Ovarian cancer remains a severe clinical challenge in China, exhibiting the highest mortality rate among gynecological malignancies, with approximately 70% of patients diagnosed at an advanced stage and a five-year survival rate dropping below 30%. The management of platinum-resistant recurrent ovarian cancer represents a historical therapeutic bottleneck, where traditional single-agent non-platinum chemotherapy yields an objective response rate of less than 10%, leaving patients with an average survival of only a few months [13]. Through the GBA measure, Mirvetuximab Soravtansine (索米妥昔单抗)—the world's first antibody-drug conjugate (ADC) targeting folate receptor alpha positive (FRα+) ovarian cancer—was successfully introduced to the mainland. By utilizing this regulatory pathway at the Sun Yat-sen Memorial Hospital of Sun Yat-sen University, 80 patients were treated with this innovative ADC by June 2025. Crucially, the GBA measure allowed these patients to access the treatment two months ahead of its national marketing authorization in China, securing a vital therapeutic window that significantly extended survival opportunities [14].

Cardiovascular Innovations: Advanced Endovascular & Electrophysiology Devices

The GBA measure has been equally impactful in the field of cardiovascular medicine, introducing state-of-the-art medical devices that minimize procedural risks and optimize patient safety:

- The EKOS Ultrasonic Assisted Thrombolysis Management System: Acute massive pulmonary embolism is a high-risk cardiovascular emergency that frequently induces rapid respiratory failure and hemodynamic shock. While traditional systemic thrombolysis carries high hemorrhagic risks and slow onset times, the EKOS system—introduced under the GBA measure—uses ultrasound waves to thin clot fibrin strands, allowing thrombolytic drugs to penetrate deeply into the thrombus. This mechanism enhances drug absorption by up to 84% within 1–2 hours, dramatically improving efficacy while reducing the required drug dosage and bleeding risks. Following its successful clinical debut at the First Affiliated Hospital of Sun Yat-sen University, local cardiovascular teams spearheaded a regional real-world study in Asian populations to accumulate vital localized evidence [14].
- Absorbable Antibacterial Envelopes (可吸收抗菌封套): In May 2022, the University of Hong Kong-Shenzhen Hospital (HKU-SZH) achieved a landmark milestone by executing the mainland's first clinical implantation of an absorbable antibacterial envelope under the GBA measure. Designed

for high-risk patients undergoing cardiac implantable electronic device (CIED) replacements—such as cross-border Hong Kong residents requiring pacemaker or implantable cardioverter-defibrillator (ICD) battery upgrades—the envelope encapsulates the device within the tissue pocket. Global multi-center real-world data (the WRAP-IT study) confirmed that this technology reduces major CIED infections by 40% and pocket-related infections by 61%. The policy allowed top-tier Hong Kong clinicians to seamlessly cross the border to co-operate on these complex surgeries, blending international clinical expertise with immediate access to cutting-edge global hardware [15].

Cumulative Public Health Milestones

The structural maturation of the GBA measure is demonstrated by its rapidly expanding operational scale. By August 2025, the program had approved a total of 45 designated medical institutions, achieving comprehensive coverage across all nine mainland cities within the GBA. The dynamically managed urgent import catalog has approved 120 distinct medical products, comprising 53 innovative pharmaceuticals—with oncology drugs constituting nearly one-third of the selections—and 67 advanced medical devices, with cardiovascular technologies representing 49% of the total hardware roster. To date, more than 13,000 patient-times have directly benefited from these rapid importations, proving that the GBA measure has successfully evolved from a local pilot into an essential, resilient mechanism for regional health equity and medical innovation [16].

7.18 Strategic Industry Insights, Opportunities, and Challenges

The implementation of the GBA measure presents a paradigm shift for the global pharmaceutical and medtech industries, offering unique strategic opportunities alongside distinct operational challenges.

Strategic Opportunities and Real-World Data (RWD): For global companies, the policy serves as a lucrative gateway, particularly for high-value items treating rare diseases or complex oncology cases [2]. Most importantly, the policy allows companies to generate crucial real-world data (RWD) [2]. Under traditional pathways, lengthy NMPA applications require extensive localized clinical trials [2]. The GBA measure allows products to be used clinically, generating localized efficacy and safety data [2]. Article 23 of the 2024 Ordinance explicitly encourages designated hospitals to conduct Real-World Studies to support formal NMPA registration applications, accelerating mainland market access [10]. Furthermore, startups can leverage the absence of mandatory registration requirements for certain devices in Hong Kong to introduce advanced technologies to the GBA with significantly reduced time and costs [2].

Challenges and Best Practices: Despite the benefits, the criteria remain exceptionally strict—products must be clinically advanced and lack equivalent domestic substitutes, limiting eligible products [2]. Furthermore, navigating complex HK drug approval processes for New Chemical Entities (NCEs) can be a time-consuming bottleneck, and vigorous competition exists with mainland pharmaceutical companies [2]. To succeed, industry experts recommend optimizing internal collaboration between Regulatory Affairs (RA) and Medical Affairs (MA), as applications require intense legal documentation synthesized with complex clinical justifications [2]. Establishing dedicated cross-border task forces comprising teams from the Mainland, Hong Kong, and Macao is highly recommended to bridge cultural and regulatory gaps and establish harmonized pharmacovigilance workflows [2]. Finally, continuous education platforms must be initiated to ensure frontline medical staff fully understand the nuanced operational constraints and legal liabilities [2].

7.19 Conclusion and Future Outlook

The Measure of using HK-registered drugs and medical devices used in HK public hospitals in Guangdong-Hong Kong-Macao Greater Bay Area (the GBA measure) stands as a monumental leap in regional healthcare integration and regulatory innovation. By meticulously delineating the criteria for designated institutions, creating a fast-tracked yet deeply rigorous application process, and establishing a watertight system of end-to-end traceability, Chinese regulatory authorities have effectively bridged the gap between disparate legal and healthcare systems.

For the 70 million residents of the GBA, this policy significantly elevates the standard of care by accelerating access to life-saving global medical innovations directly within their local mainland communities, shifting the paradigm from "waiting for medicine" to "medicine awaiting patients". Moving forward through the 15th Five-Year Plan, the continued expansion of the GBA measure—supercharged by Hong Kong's evolution into an international medical innovation hub and its implementation of an independent primary evaluation system—will rely heavily on the seamless technological integration of pre-approval databases, the robust expansion of inclusive commercial insurance networks, and the strategic utilization of real-world data. Ultimately, this synergy will profoundly benefit the entire national healthcare ecosystem and propel the realization of the "Healthy China 2030" vision.

References

1. National Development and Reform Commission, People's Government of Guangdong Province, Government of the Hong Kong Special Administrative Region, Government of the Macao Special Administrative Region. 《粤港澳大湾区发展规划纲要》 (Outline Development Plan for the Guangdong-Hong Kong-Macao Greater Bay Area). 2019.
2. Cheng CH, Cheng FY, Chow SF, Leung R, Wong J, Yau EW, Yeo A. Measure of using HK registered drugs and medical devices used in HK public hospitals in Guangdong-Hong Kong-Macao Greater Bay Area White Paper.
3. Central Committee of the Communist Party of China, State Council. 《"健康中国2030"规划纲要》 ("Healthy China 2030" Planning Outline). Beijing; 2016.
4. National Development and Reform Commission. 《"十四五"国民健康规划》 (The 14th Five-Year National Health Plan). Beijing; 2022.
5. Central Committee of the Communist Party of China. 《中共中央关于制定国民经济和社会发展第十五个五年规划的建议》 ("15th Five-Year Plan" Recommendations on Formulating the Fifteenth Five-Year

Plan for National Economic and Social Development). Xinhua News Agency; 2025.

6. Bao J. 《[健康报] 奋楫“十五五” 共绘人民健康蓝图》 (Health News: Navigating the 15th Five-Year Plan and Mapping the People’s Health Blueprint). National Health Commission of the People’s Republic of China; 2026.
7. Yau EW. Import/Export and marketing of Drugs and Medical Devices in the GBA. Hong Kong.
8. Health Bureau. 《改革藥物及醫療器械審批制度的進展以及就成立香港藥物及醫療器械監督管理中心的立法建議》 (Legislative Proposals on Reforming Drug and Medical Device Approval Systems and Establishing the Hong Kong Centre for Medical Products Regulation) (Legislative Council Paper CB(3)261/2026(03)). Legislative Council of the Hong Kong Special Administrative Region; 2026.
9. State Administration for Market Regulation, National Medical Products Administration, et al. 《粤港澳大湾区药品医疗器械监管创新发展工作方案》 (Work Plan for Regulatory Innovation and Development of Pharmaceutical and Medical Device in the Guangdong-Hong Kong-Macao Greater Bay Area) (Document No. 国市监药[2020]159号). Beijing; 2020.
10. Guangdong Provincial People's Congress Standing Committee. 《广东省粤港澳大湾区内地九市进口港澳药品医疗器械管理条例》 (Regulations on Importing Hong Kong and Macao Pharmaceuticals and Medical Devices in the Nine Mainland Cities of the Greater Bay Area). Guangzhou; 2024.
11. Guangdong Provincial Medical Products Administration, Health Commission of Guangdong Province. 《广东省粤港澳大湾区内地九市进口港澳药品申报指南》 (Guangdong Province Greater Bay Area Nine Mainland Cities Urgent Hong Kong and Macao Drug Application Guide). Guangzhou; 2024.
12. Guangdong Provincial Medical Products Administration, Health Commission of Guangdong Province. 《广东省粤港澳大湾区内地九市进口港澳医疗器械申报指南》 (Guangdong Province Greater Bay Area Nine Mainland Cities Urgent Hong Kong and Macao Medical Device Application Guide). Guangzhou; 2024.
13. Guangdong-Hong Kong-Macao Greater Bay Area Portal Network. *Typical Cases of Rules Counterpart and Mechanism Docking in the GBA (First Batch)—Case 8: “港澳藥械通” Boosts the Construction of a Healthy Bay Area*, June 2024.

14. Guangzhou Municipal People's Government. *Guangzhou Medical Expo "港澳藥械通" Special Session: Introducing High-Quality Medical Products to Solve Clinical Pain Points*, Guangzhou Daily, August 24, 2025.
15. Hong Kong Commercial Daily. *[Shenzhen-Hong Kong Dual-City Chronicles]"港澳藥械通" Adds New Cardiovascular Device Benefiting Hong Kong Patients in Shenzhen*, June 8, 2022.
16. Guangdong Provincial Medical Products Administration & Health Commission of Guangdong Province. *Regulations on Importing Hong Kong and Macao Pharmaceuticals and Medical Devices in the Nine Mainland Cities of the Greater Bay Area*, Effective December 1, 2024.

Chapter 8

Strategic Comparison: Hong Kong vs. Macao vs. Chinese Mainland (GBA Focus)

Ansel Zhang
Member, Hong Kong Local Responsible Persons Panel
ansel6399@gmail.com

8.1 Introduction

Chapters 4, 5, and 6 of this book went through the regulatory architectures of Hong Kong, Macao, and Chinese Mainland, respectively, focusing on device lifecycle management, market entrance, post-market surveillance, and the institutional evolution of regulatory authorities. Chapter 7 then zoomed out to analyze the Greater Bay Area (GBA) as an integrated policy experiment that deliberately bridges these otherwise heterogeneous systems.

This chapter builds on that foundation by answering a critical practitioner question:

Medical Device Regulation in the Guangdong-Hong Kong-Macao Greater Bay Area: A Practical Guide for Compliant and Efficient Market Entry
Edited by Jack Wong and Raymond K. Y. Tong

ISBN 978-981-5352-38-2 (Hardcover), 978-1-042-03128-3 (eBook)
www.jennystanford.com

DOI: 10.1201/9781042031283-11

Given three distinct but increasingly connected regulatory environments, how should companies strategically choose, sequence, and structure market entry across Hong Kong, Macao, and Chinese Mainland (GBA Policy)? Rather than re-describing legal provisions, this chapter offers a side-by-side strategic comparison. The objective is to help regulatory leaders translate regulatory complexity into deliberate market architecture decisions, particularly within the fast-evolving GBA context.

8.2 Comparative Regulatory Architecture

At a structural level, Hong Kong, Macao, and Chinese Mainland embody three distinct regulatory philosophies, and those differences directly shape market-entry speed, procedural predictability, compliance burden, and strategic risk tolerance.

8.2.1 Hong Kong

As discussed in Chapter 4, Hong Kong's medical device regime remains formally voluntary, but in practice it functions as a quasi-mandatory system for meaningful market participation. Its defining features include the voluntary MDACS listing framework administered by the Medical Device Division, reliance on Local Responsible Persons (LRPs) rather than direct manufacturer establishment, the absence of formal compulsory pre-market approval, and a corresponding increase in post-market surveillance, change management, and traceability obligations.

The current Hong Kong guidance demonstrates increasing alignment with internationally recognized regulatory practice, including concepts reflected in IMDRF frameworks.[1] Although the detailed design of the forthcoming mandatory medical device regime has not yet been publicly exposed, the government's longer-term regulatory roadmap has made its direction clear. At the same time, the current Medical Device Division (MDD) and the future Centre for the Medical Products Regulation (CMPR) have continued

[1]Department of Health, Medical Device Division. (2025). *Guidance notes: GN-01 – Overview of the medical device administrative control system.* Hong Kong SAR Government.

to engage closely with local industry stakeholders while expanding cooperation with overseas regulators, including the signing of a memorandum of understanding with Singapore's Health Sciences Authority in August 2025.[2]

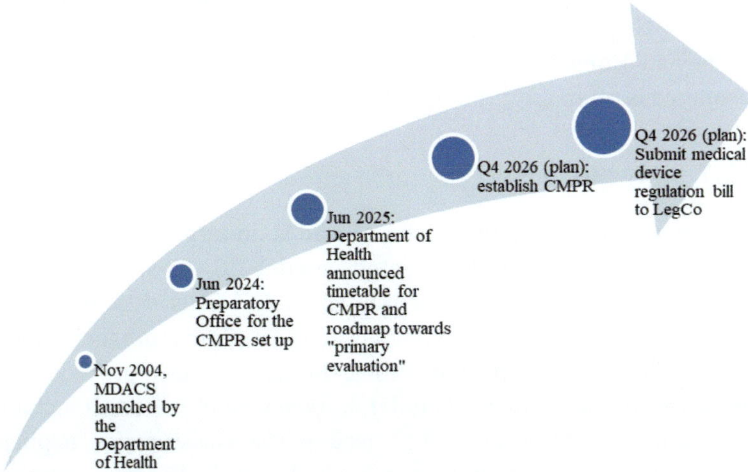

Figure 8.1 Milestone of Hong Kong medical device regulation.

As of the time of writing, no public exposure draft has been issued setting out the detailed requirements of the upcoming mandatory regime. Nevertheless, health authority communications have repeatedly emphasized the value of early participation under the existing MDACS framework, both to maximize the current cost advantage of voluntary listing and to support a smoother transition to future mandatory regulation. Industry response has been correspondingly active, as reflected in Local Responsible Person panel discussions and in the continued growth of the MDACS database,[3] which now records more than 9,900 listed medical devices in Hong Kong.

Because of Hong Kong's health system structure, public hospitals are managed by the Hospital Authority, a separate statutory body that operates alongside the Medical Device Division under the

[2]Health Sciences Authority. (2025, August 13). Health Sciences Authority signs memorandum of understanding with Hong Kong Department of Health to enhance healthcare regulatory cooperation.

[3]Department of Health, Medical Device Division. (n.d.). Medical device administrative control system – List of medical devices. Hong Kong SAR Government.

broader public health framework. Based on publicly available Hospital Authority tender requirements, certain product categories are subject to additional technical expectations beyond device listing. Given that public hospitals are a critical customer segment for many medical device companies, this underscores that medical device listing is only the starting point of the commercialization journey in Hong Kong, and that further investment of effort and resources is often required in the next stage.

8.2.2 Macao

As discussed in Chapters 5 and 6, Macao is undergoing a deliberate transition from a trade-oriented model to a science-driven regulatory system under Law No. 12/2025.[4] This shift positions Macao as an intermediate model within the region: more formalized and legally structured than Hong Kong, yet less procedurally burdensome than Chinese Mainland. Its key characteristics include mandatory registration or filing based on risk classification, explicit clinical evaluation requirements for higher-risk devices, licensing requirements for manufacturing and operational activities, and transitional arrangements for existing products designed to reduce disruption to local supply and clinical use.

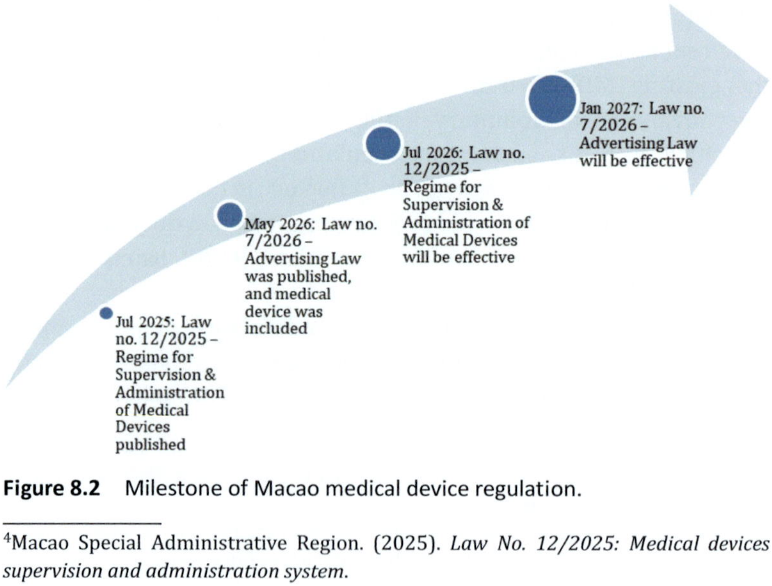

Figure 8.2 Milestone of Macao medical device regulation.

[4]Macao Special Administrative Region. (2025). *Law No. 12/2025: Medical devices supervision and administration system.*

Unlike Hong Kong, Macao does not provide a voluntary registration phase before the introduction of formal regulation. As a result, the transition from an effectively unregulated environment to a regulated medical device system is likely to have a more significant impact on industry participants and end users. Based on the published law and available drafting guidance, the health authority has nevertheless sought to smooth that transition and reduce direct and indirect social disruption through a series of relief measures, including a three-year grace period, selected licensing exemptions for low-risk medical device retailers, and selected exemptions for existing medicine trade operators.

Like many health authorities, Macao has gradually expanded its regulatory focus from medicines alone to the broader category of medical products, including medical devices. As a result, its medical device framework still occasionally reflects its historical roots in pharmaceutical regulation. One example is the revision of the Macao Advertising Law[5] in May 2026, which brought medical devices within the scope of advertising control for the first time. Under that framework, pre-notification for medical device advertising can be understood as a lower-intensity control mechanism analogous to the pre-approval model traditionally applied to medicine advertisements. This development illustrates how the health authority is actively calibrating its level of control over medical devices while drawing on prior regulatory practice in the pharmaceutical sector. At the same time, Macao's regulatory trajectory also reflects increasing globalization and regulatory reliance, as demonstrated by its official admission to the Global Harmonization Working Party (GHWP) at the 28th Annual Meeting in Kuala Lumpur in December 2024.[6]

8.2.3 Chinese Mainland—GBA Policy

Chinese Mainland's NMPA system represents the highest regulatory threshold among the three jurisdictions, particularly for Class II and Class III devices. It is characterized by mandatory registration supported by extensive technical dossiers, frequent local clinical evidence expectations, intensive inspection and enforcement

[5]Macao Special Administrative Region. (2026, May 11). *Law No. 7/2026: Advertising law (Lei da publicidade).*

[6]Government of Macao Special Administrative Region. (2024, December 13). 藥物監督管理局參加全球醫療器械法規協調會年會 中國澳門獲通過成為協調會成員.

mechanisms, and substantially longer review timelines with higher sunk compliance costs. Within the GBA Measure framework, selected products may access Chinese Mainland through pathways that do not require full conventional NMPA registration, but only under narrowly defined conditions. In this context, Hong Kong and Macao approvals are becoming increasingly important as policy gateways rather than substitutes for the Chinese Mainland medical system.

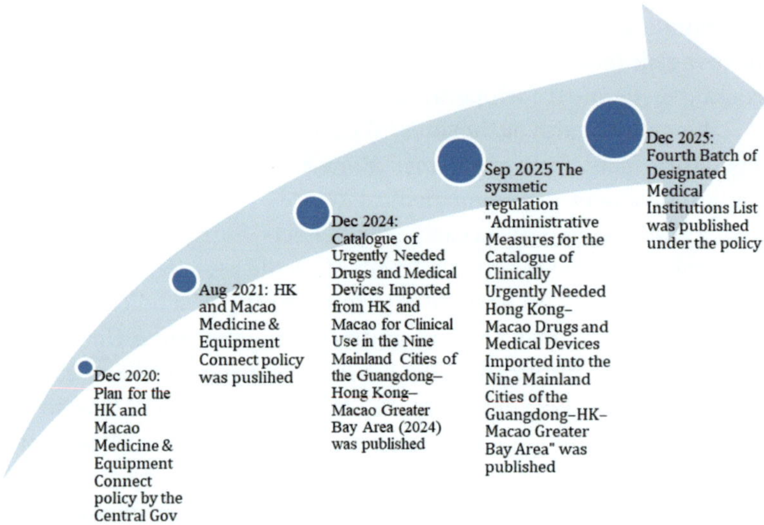

Figure 8.3 Milestone of Chinese Mainland GBA medical device regulation.

The policy was first planned in December 2020 and initially launched in 2021 at a single hospital in one Chinese Mainland city. As the framework continued to evolve and expand, by the end of 2025 a total of 71 hospitals[7] across nine Chinese Mainland cities had been designated under the policy. From the medical device perspective, there are currently 37 active devices from 20 companies qualified under the framework. It should be noted that the qualified device list is dynamic, as products may be removed for various reasons, including where the device subsequently obtains formal NMPA approval through the standard registration or filing pathway.

[7]广东省卫生健康委员会 & 广东省药品监督管理局. (2025).广东省卫生健康委员会 广东省药品监督管理局关于印发粤港澳大湾区“港澳药械通”第四批指定医疗机构名单的通知

As defined in the Administrative Measures for the Catalogue of Clinically Urgently Needed Hong Kong-Macao Drugs and Medical Devices Imported into the Nine Mainland Cities of the Guangdong-Hong Kong-Macao Greater Bay Area,[8] the policy targets "medical devices that meet urgent clinical needs and possess advanced clinical application profiles." Based on the latest Catalogue of Urgently Needed Drugs and Medical Devices Imported from Hong Kong and Macao for Clinical Use in the Nine Mainland Cities of the Guangdong-Hong Kong-Macao Greater Bay Area (2024),[9] the data show that medical devices under the GBA policy are concentrated in a limited number of disease areas.

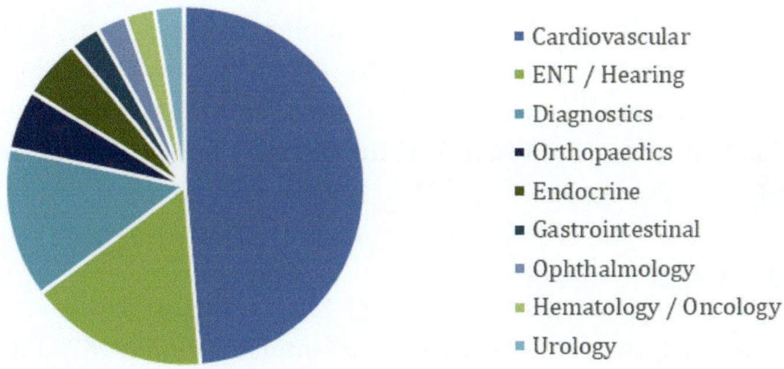

Figure 8.4 GBA policy device quantity by disease area.

8.3 Comparison

8.3.1 Time and Cost

Hong Kong's health authority does not publish an official target review timeline for medical device applications. However, the current framework provides an Expedited Approval Scheme[10] applicable to all types of medical devices. At present, listing under MDACS remains free of charge.

[8]广东省药品监督管理局, & 广东省卫生健康委员会. (2025). 广东省粤港澳大湾区内地九市临床急需进口港澳药品医疗器械目录管理办法.

[9]广东省药品监督管理局, & 广东省卫生健康委员会. (2024). 粤港澳大湾区内地九市临床急需进口港澳药品医疗器械目录（*2024*年

[10]Hong Kong Department of Health, Medical Device Division. (2025). Notes for the expedited approval scheme for medical device listing applications.

In Macao, the health authority plans to establish differentiated review timelines based on product classification. Priority review may be granted to medical devices intended to address rare diseases or malignant tumors with clear clinical advantages, meet unmet diagnostic or treatment needs in elderly populations, are specifically designed for pediatric use with demonstrated benefits, fill urgent clinical gaps where no alternatives exist, demonstrate significant innovation, or are otherwise designated by ISAF as eligible for priority review. Filing of Class I and Class IIa medical devices will be free of charge, whereas registration of Class IIb and Class III medical devices will require payment of a registration fee. However, ISAF has not yet published the exact fee amounts.

For the Chinese Mainland GBA policy pathway, the process generally requires more time and effort because case-by-case communication and expert meetings are often necessary. At present, applications under the Chinese Mainland GBA policy remain free of charge.

8.3.2 Local Representatives

In Hong Kong, the LRP is the key local interface under MDACS. The LRP maintains the device listing, serves as the contact point for the health authority, and coordinates complaint handling, adverse incident reporting, and other post-market actions. Although Hong Kong does not require a formal marketing authorization holder, the LRP effectively anchors local market access and lifecycle management. Because the LRP controls core regulatory interfaces, the main structuring issue is governance rather than formal license ownership. If dossier access, vigilance duties, complaint records, and transfer rights are not clearly defined, external LRP arrangements can create dependency and transition risk. For higher-risk or strategically important products, companies often prefer in-house LRPs or tightly governed structures.

In Macao, accountability is expected to rest more directly with the filing or registration holder under the emerging medical device regime. That local holder is expected to maintain control over the product dossier, ensure compliance with applicable filing or registration requirements, support regulatory communications, and oversee post-market obligations within the local market.

Distributors may still play an important commercial role, but the regulatory model places greater emphasis on the holder's own ability to manage the product lifecycle and maintain compliance in line with the expectations of the medical products regulator. This means Macao is less suited to a loosely controlled distributor-led model where the manufacturer retains limited visibility over regulatory execution. Instead, the framework tends to favor stronger local establishment, closer distributor oversight, and tighter alignment between the holder's quality system and local regulatory expectations. From a strategic perspective, a well-structured Macao holder can support not only local commercialization, but also a stronger regional compliance profile within the Greater Bay Area.

In Chinese Mainland, local responsibility is distributed across multiple actors rather than concentrated in a single representative. Depending on the pathway, these may include the marketing authorization holder, local agents, distributors, and, under the Greater Bay Area policy, designated enterprises linked to approved hospitals and import arrangements. Together, these stakeholders are responsible for regulatory communication, import and distribution control, traceability, complaint and adverse event handling, and ongoing compliance with inspection and post-market requirements.

The key difference across the three markets lies in where local control is placed. Hong Kong concentrates operational responsibility in the LRP, making contractual governance and transfer rights critical. Macao places greater weight on the filing or registration holder's direct control over lifecycle compliance, making local entity capability more important. Chinese Mainland, especially under the GBA pathway, relies on a multi-actor structure in which hospitals, designated enterprises, agents, and other parties must operate in a coordinated and traceable manner. Accordingly, Hong Kong presents the greatest dependency risk on a single local representative, Macao demands stronger holder-led compliance capability, and Chinese Mainland requires the most complex delegated governance model.

8.3.3 Discussion

The cost-risk trade-off across the three jurisdictions is not merely a question of registration fees or dossier volume; it reflects fundamentally different allocations of legal exposure, commercial flexibility, and long-term strategic value. In Hong Kong, the low upfront cost and relatively rapid access associated with the MDACS framework can be highly attractive, particularly for companies seeking early market presence, initial user feedback, or participation in public procurement and Greater Bay Area access mechanisms. However, that lower entry threshold is offset by other forms of risk, including dependence on Local Responsible Person arrangements, comparatively limited pre-market regulatory shielding if post-market incidents arise, and potential reputational or operational disruption as the system continues evolving toward a more formal mandatory regime.

By contrast, Chinese Mainland requires substantially greater capital, documentation, and organizational commitment at the outset, yet that investment may yield stronger legal certainty, broader institutional acceptance, and more durable long-term market positioning once approval is secured. The system is therefore better suited to products or portfolios for which companies are prepared to absorb high evidence burdens in exchange for national-scale commercialization opportunities.

8.4 Strategy

For startups and emerging innovators, Greater Bay Area regulatory strategy is closely tied to capital efficiency, evidence generation, and survival. A staged approach is often more practical than an immediate full-scale expansion. Hong Kong can serve as an initial launch market because MDACS allows relatively fast entry, supports early product-market validation, and helps generate clinical or commercial credibility with limited upfront cost. Macao may then function as a selective second step for more mature products, offering stronger regulatory validation and a more structured compliance environment. For some technologies, GBA access mechanisms can further support data generation before full Chinese Mainland entry. By contrast, direct NMPA registration is often deferred, limited to priority products, or pursued through licensing, distribution, or strategic partnerships when capital and internal capabilities are constrained. That flexibility does not remove structural risk. Startups often rely on external Local Responsible Persons, distributors, or service providers, creating governance vulnerabilities if contractual rights, data access, complaint handling, or transfer arrangements are weak. Companies using Hong Kong as an initial proving ground should therefore protect listing ownership, post-market responsibilities, and future transition pathways from the outset. Otherwise, an efficient early strategy may become difficult to scale, transfer, or defend in due diligence, investor review, or acquisition discussions. The strongest startup strategy is therefore not the cheapest, but the one that preserves optionality while building credible regulatory assets over time.

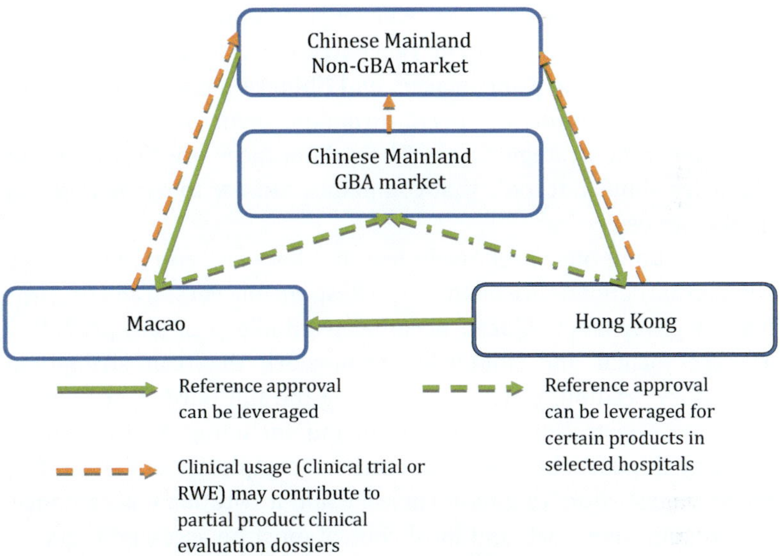

Figure 8.5 Launch strategy across different jurisdictions.

Large multinational corporations approach the region differently. With dedicated regulatory teams, mature quality and vigilance systems, and longer investment horizons, they typically use Hong Kong and Macao not as isolated markets but as parts of a broader regional portfolio strategy. Hong Kong can support lower-risk or legacy products, Macao can provide stronger formal positioning without the full burden of Chinese Mainland registration, and NMPA entry can be reserved for flagship or strategically important products. This portfolio-tiering approach helps multinational companies spread regulatory exposure rather than concentrate it.

By allocating products across jurisdictions based on risk, lifecycle stage, and commercial value, they can improve regulatory return on investment and reduce bottlenecks to regional growth. The advantage of scale is therefore not just the ability to register more products, but the ability to build a more resilient and risk-adjusted regional regulatory architecture.

8.5 Conclusion

This chapter shows that regulatory strategy in the Greater Bay Area is no longer a simple choice between jurisdictions, but a question of how to design and sequence market entry across different regulatory architectures. Hong Kong, Macao, and Chinese Mainland each offer distinct combinations of speed, formality, governance complexity, and long-term strategic value. Their differences are therefore not obstacles alone, but tools that companies can use to shape regional market access.

Hong Kong offers relatively fast and low-cost entry, but places substantial dependence on the Local Responsible Person and on post-market governance. Macao occupies a middle position, providing a more formal and holder-led framework that can strengthen regulatory credibility without reaching the full burden of Chinese Mainland registration. Chinese Mainland, including the GBA policy pathway, imposes the highest operational complexity, but also offers the strongest route to long-term scale and institutional acceptance. As a result, time, cost, and local control must be assessed together rather than in isolation.

Part IV

Outlook and Action Plan

Chapter 9

Future Outlook and 90-Day Action Plan

Jack Wong[a] and Zhuo Chen[b]
[a]*Asia Regulatory Professionals Association*
[b]*Former China Regulator*
jackwong@nus.edu.sg

Medical device regulation in Hong Kong, Macao, and other markets inside the Greater Bay Area (GBA) is shifting from partially voluntary, fragmented systems to a more integrated, risk-based, and enforceable framework. This chapter looks ahead over the next three to five years and then sets out a practical 90-day action plan to help companies respond proactively.

9.1 Need Action Now

Over the next three to five years, Hong Kong, Macao, and the Greater Bay Area will become significantly more demanding and strategic for medical device manufacturers, as voluntary schemes evolve into mandatory regimes, enforcement tightens, and GBA integration makes local approvals a potential gateway into Mainland institutions.

Medical Device Regulation in the Guangdong-Hong Kong-Macao Greater Bay Area: A Practical Guide for Compliant and Efficient Market Entry
Edited by Jack Wong and Raymond K. Y. Tong

ISBN 978-981-5352-38-2 (Hardcover), 978-1-042-03128-3 (eBook)
www.jennystanford.com

DOI: 10.1201/9781042031283-13

At the same time, authorities are raising expectations for evidence and post-market surveillance, which means that companies treating these markets seriously—by investing early in systems, documentation, and strong partners—can convert regulatory readiness into a real competitive edge. For a concise overview of these dynamics, and how they reinforce each other, see Table 9.1.

Table 9.1 Reasons why the next three to five years matter

No.	Reason	Brief Explanation
1	Shift to mandatory regimes	Voluntary or semi-formal schemes in Hong Kong and Macao are moving towards clearer statutory requirements, so 'light-touch' compliance will no longer be sufficient.
2	Stronger enforcement	Authorities are gaining more tools and resources to inspect, sanction, and publicise non-compliance, increasing reputational and business risk.
3	GBA integration	New Greater Bay Area mechanisms make Hong Kong and Macao approvals more strategically valuable as potential gateways into selected Mainland institutions.
4	Higher evidence and PMS expectations	Documentation, clinical evidence, and post-market surveillance standards are rising, especially for higher-risk and innovative devices.
5	Competitive differentiation	Early movers who upgrade systems and partners now can turn compliance into a commercial advantage in tenders, hospital access, and partnerships.

9.2 Hong Kong: From Voluntary to Statutory

Hong Kong is moving from the voluntary Medical Device Administrative Control System (MDACS) towards a statutory regime anchored in primary legislation and supported by detailed regulations and guidance. This transition will progressively alter how products are brought to market, maintained, and monitored.

9.2.1 Likely Direction of the Statutory Regime

The emerging regime is expected to include mandatory registration for defined device categories, particularly medium- and high-risk devices and those used in public institutions. Legal obligations for Local Responsible Persons (LRPs) and manufacturers will become clearer, supported by investigation powers and penalties. Post-market surveillance (PMS) and vigilance will be more structured, with strengthened expectations for incident reporting and Field Safety Corrective Actions.

9.2.2 Key Areas of Regulatory Emphasis

Regulators in Hong Kong are likely to focus on four areas: clinical evidence for higher-risk and innovative devices, the robustness of quality management systems at both manufacturer and LRP level, the timeliness and quality of vigilance reporting, and the integrity of supply chains. It is prudent to prepare dossiers and internal systems now with these expectations in mind, rather than only meeting the current MDACS minimum. See Table 9.2 for a detailed comparison.

Table 9.2 Hong Kong regulatory trajectory

Phase	Regulatory Focus	Implication for Companies
Voluntary MDACS	Listing encouraged but not mandated	Low barrier to entry; limited oversight
Procurement leverage	MDACS listing linked to public tenders	Listing becomes commercially essential
Statutory regime	Mandatory registration, clearer duties	Higher compliance bar, stronger PMS
Integrated future	Alignment with GBA initiatives	HK approvals support wider regional access

Practical example: A mid-sized orthopaedics company treated MDACS listing as optional and only registered a few flagship implants. When a major public tender required MDACS listing for all high-volume items, it had to rush multiple dossiers at once, accept sub-optimal labelling, and rely heavily on its LRP. A competitor that had quietly listed its broader portfolio two years earlier faced no disruption and won a larger share of the tender.

9.3 Macao: Consolidating Law 12/2025

Macao's Law 12/2025 establishes a comprehensive framework for classification, registration, listing, licensing, and supervision of medical devices. The coming years will focus on implementing and operationalising this law.

9.3.1 Implementation and Transitional Dynamics

Following promulgation of Law 12/2025, the authorities will issue implementing regulations, standard forms, and technical guidance to clarify procedures. Early practice often includes transitional arrangements for products already on the market, but the scope for flexibility narrows over time as systems stabilise. Companies should use the initial period to regularise portfolios, documentation, and licences rather than postponing action.

9.3.2 Enforcement Pattern and Business Plan

Enforcement usually moves from education and voluntary corrections, through targeted warnings and fines, to more visible sanctions and licence suspensions in serious cases. Firms that ignore early signals or delay compliance efforts risk becoming high-profile examples when authorities demonstrate the effectiveness of the new framework. See Table 9.3 for the typical post-enactment pattern.

Table 9.3 Typical post-enactment pattern after Law 12/2025

Year after Law	Regulatory Focus	Recommended Company Actions
0–1	Roll-out, FAQs, basic supervision	Map products; prioritise registrations and licences
1–2	Targeted inspections, initial penalties	Close priority gaps in documentation and labelling
2–3	Routine inspections and monitoring	Optimise processes, reduce recurring deficiencies
3+	Mature, data-driven enforcement	Treat Macao as a fully mature regulatory market

In practice, a prepared company in Macao by year two will have:

- *A clean list of all devices on the market, mapped to the new classification rules.*
- *Contracts updated so licence holders clearly own vigilance and reporting duties.*
- *At least one internal audit completed against Law 12/2025 requirements.*

9.4 Greater Bay Area: Integration and Pilots

The GBA policy framework aims to create an integrated innovation hub connecting Hong Kong, Macao, and Mainland cities such as Shenzhen and Guangzhou. For medical devices, this includes mechanisms that allow certain Hong Kong-approved products to be used in designated Mainland healthcare institutions.

9.4.1 Strategic Role of Hong Kong and Macao in GBA

Under specific GBA arrangements, drugs and medical devices that are approved or used in Hong Kong may be introduced into designated Mainland institutions in the GBA for urgent clinical need or innovation pilots. This effectively positions Hong Kong approvals, and to a lesser extent Macao status, as potential gateways into parts of the Mainland market for selected products.

9.4.2 Expected GBA Trends

In the coming years, several trends are likely to strengthen the strategic role of Hong Kong and Macao within the Greater Bay Area innovation framework:

- **Expansion of pilot programmes**: More designated Mainland healthcare institutions will be authorised to procure and use Hong Kong-approved devices for urgent clinical need or innovation projects.
- **Refined eligibility criteria**: Authorities will clarify which device categories, risk classes, and clinical indications qualify for cross-border mechanisms, focusing on unmet need and innovation.

- **Digital tracking systems**: Enhanced electronic systems will monitor cross-border movement, device utilisation, and clinical outcomes, improving traceability and regulatory confidence.
- **Data-sharing frameworks**: Real-world performance data from Hong Kong use may be considered in Mainland benefit-risk assessments or post-market reviews for selected products.

Impact on manufacturers: Companies that plan their Hong Kong submissions with GBA opportunities in mind—rather than viewing Hong Kong purely as a local market—will be structurally better positioned to access multi-city networks of leading hospitals and to leverage Hong Kong approvals as evidence in broader regional strategies. Figure 9.1 illustrates how manufacturers can use Hong Kong and Macao as dual gateways, with Hong Kong approvals opening routes to both local hospitals and GBA pilot institutions, while Macao provides a parallel pathway into its own healthcare system.

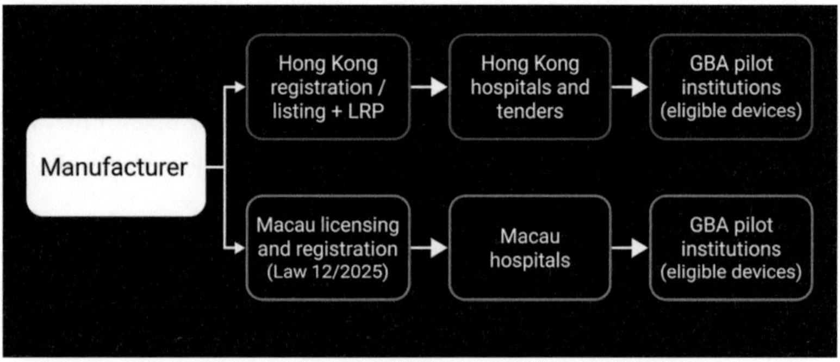

Figure 9.1 Schematic of Hong Kong and Macao as GBA gateways.

Illustrative scenario: *A diagnostics manufacturer plans its Hong Kong submission for a novel cardiac biomarker assay. Instead of treating Hong Kong as a small standalone market, it aligns the clinical evidence package with Mainland expectations, chooses a Hong Kong hospital already participating in GBA pilots, and documents real-world performance from that site. When a suitable GBA mechanism opens, the company can re-use most of its data and local relationships, shortening the path into a leading Mainland cardiac centre by more than a year.*

9.5 Emerging Topics: Digital Health, Software, and AI

Digital health products, standalone software, and AI-enabled devices are challenging traditional regulatory categories. Hong Kong and Macao are expected to draw on global approaches when clarifying their own requirements.

9.5.1 Classification and Evidence Expectations

Authorities are likely to refine criteria for when software constitutes a medical device, emphasising intended purpose, risk, and impact on clinical decisions. Evidence expectations may expand to include real-world performance, cybersecurity, and algorithm transparency, particularly for AI-based decision support tools. Developers should anticipate these themes in their design and documentation strategies. See Table 9.4 for recent key AI/ML regulations and guidance for medical devices.

Practical tips for AI developers

- *Build an internal register of all models in use, their versions, and intended clinical roles.*
- *Pre-define what counts as a "minor" versus "major" software update from a regulatory point of view.*
- *Keep a plain-language "AI factsheet" for doctors and hospitals that explains what the tool does, its limits, and how it is monitored.*

9.5.2 Post-Market Surveillance for Digital Products

For software and AI, post-market surveillance may involve version-specific tracking, impact assessment of updates, and monitoring of data quality and model performance over time. Companies will need processes to demonstrate that updates do not compromise safety or effectiveness and that any emerging issues are detected and managed promptly.

Table 9.4 Recent key AI/ML regulations and guidance for medical devices

Region	Key Recent Documents (AI/ML & SaMD)	Main Focus Areas	Implications for Manufacturers
United States (FDA)	AI/ML-Based SaMD Action Plan and web hub on AI in SaMD (updated 2025); Final guidance on Predetermined Change Control Plans (PCCPs) for AI-enabled device software functions (Dec 2024); Good Machine Learning Practice (GMLP) and transparency guiding principles for ML-enabled medical devices (2024–2025).	Lifecycle-based control of algorithm changes via PCCPs; expectations for data quality, training/validation, and human oversight; transparency to users and regulators.	Need to pre-define allowable AI updates in PCCPs; build robust ML development and monitoring processes; provide clear user information on AI behaviour, limitations, and update policies.
European Union	EU Artificial Intelligence Act (Regulation 2024/1689); MDCG 2025-6 joint guidance on MDR/IVDR and AI Act interaction (drafted 2025).	Treats AI components of MDR/IVDR-regulated devices as 'high-risk' with extra requirements on data governance, technical robustness, human oversight, and post-market monitoring; conformity assessment remains under MDR notified bodies.	Must integrate AI Act obligations into existing MDR/IVDR processes; document dataset quality and bias control; manage tension between adaptive AI and MDR change-control expectations.

Region	Key Recent Documents (AI/ML & SaMD)	Main Focus Areas	Implications for Manufacturers
China (NMPA)	Guideline on Artificial Intelligence Medical Devices (draft guidance)—scope and lifecycle expectations for AI devices (Class II–III); 2023 'Medical Device Design and Development Guidelines No. 38' specifying clinical evaluation metrics for AI diagnostic software; 2024–2025 measures on high-end devices including AI systems, emphasising whole-lifecycle regulation and simplified change registration when core algorithms unchanged.	Classifies most AI medical software as Class III; strict pre-market review; prescriptive clinical performance metrics (e.g., AUC, F1 score) and sample-size expectations; streamlined post-market optimisation when core model is stable.	Need early alignment with NMPA on study design and metrics; plan for more intensive initial submissions; leverage simplified procedures for updates that do not alter core algorithms.
Japan (PMDA/MHLW)	Priority review scheme for 'Program Medical Devices'/ SaMD, including innovative AI software (pilot from 2022; formalised from 2023).	Faster review (target ~6 months) for innovative SaMD/AI products meeting specific criteria; emphasis on total-product-lifecycle management, post-market data collection, and cybersecurity for connected software.	Opportunity for accelerated access if AI product is first-in-Japan or globally novel; need structured real-world data plans and continuous evidence generation to support staged approvals and renewals.

Example: *A radiology AI vendor agrees with its Hong Kong LRP that every quarterly model update will be accompanied by a short impact report (datasets used, key performance indicators, and any new failure modes). When a potential bias issue is raised by a hospital, the team can quickly show which version was in use, its validation data, and the corrective steps taken, turning a potential crisis into a demonstration of good governance.*

9.6 Building a Hong Kong-Macao-GBA Regulatory Playbook

A structured internal playbook can help organisations handle evolving requirements in a consistent, efficient way across Hong Kong, Macao, and relevant GBA activities. The playbook should be a living document aligned with corporate strategy and risk appetite.

9.6.1 Core Components of the Playbook

Key components typically include regulatory pathway strategies, criteria for choosing LRPs and distributors, change and lifecycle management rules, PMS workflows, and standard documentation templates. These components should be tailored by device category and risk class to avoid over- or under-engineering processes. See Table 9.5 for example contents.

Table 9.5 Example contents of a regional regulatory playbook

Section	Content Highlights
Strategy	Market priorities, portfolio tiers, risk appetite
Pathways	Flowcharts for HK, Macao, and GBA-related routes
Partners	LRP and distributor selection criteria and KPIs
Lifecycle	Change categorisation and submission triggers
PMS	Complaint handling, vigilance, FSCA workflows
Documentation	Dossier templates and minimum evidence sets

9.7 Overview of the 90-Day Action Plan

The remainder of this chapter sets out a 90-day plan aimed at moving an organisation from a reactive posture to a proactive and structured approach in Hong Kong, Macao, and GBA.

9.7.1 Four-Phase Structure

The plan has four phases: assessment and mapping (days 1–30), design and decision-making (days 31–60), implementation and communication (days 61–90), and continuous improvement beyond day 90. Each phase has specific deliverables and is designed to be achievable for a medium-sized manufacturer with existing activities in Asia. See Table 9.6 for the indicative timeline.

Table 9.6 Indicative 90-day timeline

Activity	Timeline
Gap assessment	days 1–30
Playbook design	days 31–60
Process and contract implementation	days 61–90
Monitoring	from day 60 onwards

9.8 Days 1–30: Assessment and Mapping

The first 30 days focus on understanding the current situation across products, partners, and processes.

9.8.1 Product and Activity Inventory

Companies should create a consolidated inventory listing all devices marketed, supplied, or planned for Hong Kong, Macao, and GBA-related activities. For each device, the inventory should record classification, intended use, regulatory status, and key customers or tenders.

Checklist: *In practice, a workable 30-day inventory need only capture: (1) product name and model, (2) current regulatory status in Hong*

Kong, Macao, and GBA-related use, (3) key customers or tenders, and (4) who is the LRP or licence holder. Anything more detailed can be added later.

9.8.2 Partner and Contract Review

A parallel exercise should catalogue Local Responsible Persons in Hong Kong, distributors and licence holders in Macao, and any partners linked to GBA institutions. Contractual arrangements should be reviewed to confirm allocation of regulatory responsibilities, including documentation maintenance, vigilance, change notification, and interactions with authorities.

Tips: *Another important contract is tender commitment with hospitals, such as the Hospital Authority in Hong Kong. The contract could carry commitment on continued supply, and hence the company needs to ensure the registration is completed and valid during the transition phase.*

9.8.3 High-Level Gap Assessment

Using information from the inventory and partner review, teams can prepare a high-level gap assessment, for example using a red–amber–green scale, covering documentation, licences, PMS, partner performance, and internal awareness. This need not be exhaustive but should be sufficient to prioritise subsequent work. See Table 9.7 for an example snapshot.

Table 9.7 Example 30-day gap snapshot

Area	Hong Kong	Macao	GBA-Related
Product inventory completeness	Amber	Red	Amber
Dossier readiness	Amber	Red	Amber
Partner capability and contracts	Amber	Amber	Red
PMS records and workflows	Amber	Red	Amber
Management awareness	Red	Red	Red

9.9 Days 31–60: Design and Decision-Making

The second phase translates the assessment into a structured strategy and playbook.

9.9.1 Market and Portfolio Prioritisation

Based on commercial priorities and regulatory gaps, companies should decide which products are strategic in Hong Kong, Macao, and GBA over the next three to five years. For these products, investing in comprehensive dossiers, strong partners, and rigorous PMS is justified, while less strategic products may be managed with a lighter approach or phased out.

9.9.2 Playbook Development

Teams should draft or refine the regional regulatory playbook using the structure described earlier. This includes developing pathway flowcharts, defining criteria for categorising changes, and assigning decision-making responsibilities across global and local functions.

9.9.3 Partner and Contract Strategy

This period is also suitable for evaluating whether current LRPs and distributors are fit for future requirements. If gaps are identified, companies can plan for partner changes, contract renegotiations, or internal capability-building to reduce reliance on external parties for critical compliance activities.

9.10 Days 61–90: Implementation and Communication

The third phase focuses on executing priority actions and embedding new ways of working.

9.10.1 Priority Process and Dossier Upgrades

For Hong Kong, efforts may focus on dossier enhancement for high-priority devices, strengthening change-control workflows in line with guidance on modifications to listed devices, and aligning LRPs with the new expectations. For Macao, this period can be used to initiate or complete key licence applications and registrations or listings required under Law 12/2025. For GBA, at least one flagship product pathway should be defined, where feasible.

Example of a 60–90 day win: *One global company used this phase to standardise its Hong Kong change-control checklist onto a single A4 form and trained all regional teams on it. Within six months, the number of "surprise" changes discovered during audits fell sharply, and discussions with LRPs became more fact-based and less reactive.*

9.10.2 Training and Stakeholder Engagement

Internal regulatory and quality staff, commercial teams, and external partners should receive focused training on the new playbook, including practical scenarios and escalation pathways. Training should emphasise what actions to take when specific events occur, rather than only describing legal requirements in abstract terms.

Practice tip: *At the start of the 90-day plan, ask senior management to nominate a single "decision owner" for Hong Kong-Macao-GBA. This does not replace local experts but avoids stalemates when global and regional teams disagree on priorities.*

9.10.3 Management Communication and Roadmap

Before the 90-day period ends, regulatory teams should brief senior management on key risks and opportunities identified, actions completed, and a proposed 12-month roadmap. This helps secure continued support and resources, and positions Hong Kong, Macao, and GBA as strategic components of the company's broader market-access plan.

9.11 Beyond Day 90: Continuous Monitoring and Adaptation

Regulatory change will continue beyond the initial 90-day plan, particularly as Hong Kong finalises statutory arrangements and Macao fully implements Law 12/2025.

9.11.1 Monitoring and Internal Communication

Responsibility should be assigned for monitoring regulatory updates in Hong Kong and Macao and for tracking new GBA initiatives related to medical devices. A short internal bulletin issued quarterly can keep stakeholders informed without overloading them.

Simple practice: *Rather than long newsletters, one regulatory team in Asia sends a one-page quarterly "HK–Macao–GBA update" to commercial and quality colleagues. It uses three headings only: "What has changed", "What we are doing about it", and "What you need to do". This format keeps busy teams engaged and reduces email fatigue.*

9.11.2 Periodic Playbook Review

The regional playbook should be reviewed at least annually to incorporate new laws, guidance, and organisational experience. Lessons from audits, inspections, significant submissions, and incident management should be reflected in updated decision rules and templates.

9.12 Conclusion

The regulatory landscape in Hong Kong, Macao, and the GBA is moving towards greater structure, higher expectations, and deeper integration. Companies that respond systematically—through careful assessment, a clear 90-day action plan, and a living regional playbook—can transform regulatory preparedness into a competitive advantage in this evolving environment.

Practice tip: *During the early years of Law 12/2025, schedule at least one informal visit or call with the authority (via your local partner) framed as a "clarification discussion", not a "negotiation". Regulators tend to look favourably on companies who ask intelligent questions early rather than pleading for exceptions later. Joining Hong Kong LRP Panel meeting in Hong Kong, and Guangdong–Hong Kong–Macao Greater Bay Area Regulatory Science Training Programme 3+1: Workshop 1 in Macao could be the great choice.*

Let's support and participate in the Medical Device Regulatory Initiatives in the Greater Bay Area

By engaging through this network and participating in workshops organised by the regulators, universities and stakeholders can stay informed and contribute to shaping the future of medical device regulation in the Greater Bay Area.

Index